Raman Thermometry

Yanan Yue

Raman Thermometry

For Micro/Nanoscale Characterization

Springer

Yanan Yue
Department of Mechanical and Manufacturing Engineering
Miami University
Oxford, OH, USA

ISBN 978-981-95-5984-8 ISBN 978-981-95-5985-5 (eBook)
https://doi.org/10.1007/978-981-95-5985-5

This Springer imprint is published by the registered company Springer Nature Singapore Pte Ltd.
The registered company address is: 152 Beach Road, #21-01/04 Gateway East, Singapore 189721, Singapore

Preface

The continuous miniaturization of electronic and photonic devices, along with the rapid development of nanotechnology, has brought unprecedented challenges to thermal management and energy transport at micro- and nanoscale dimensions. Accurate and non-invasive temperature measurement techniques are crucial for understanding thermal behaviour, guiding material design, and optimizing device performance at these scales. Among various techniques, Raman thermometry has emerged as a powerful and versatile tool due to its high spatial resolution, contactless nature, and sensitivity to temperature-dependent phonon behaviour.

This book, *Raman Thermometry—For Micro/Nanoscale Characterization*, aims to provide a comprehensive and systematic overview of the principles, techniques, and applications of Raman-based thermometry methods in micro- and nanoscale systems. It is intended for researchers, graduate students, and engineers in the fields of thermal sciences, nanotechnology, materials science, and applied physics.

Beginning with an introduction to thermophysical properties at small scales, the book explores the fundamental physics of heat transport and various electro-thermal and opto-thermal characterization methods. Subsequent chapters explore the physical principles of Raman thermometry, covering both steady-state and transient techniques, as well as specialized applications such as interfacial heat transfer, convective processes, phonon non-equilibrium, and near-field heat transport. Other topics like tip-enhanced Raman thermometry and its ability to probe ballistic heat transport are also discussed in depth.

By integrating recent research findings with theoretical works and experimental methodologies, this book not only highlights the current state of the art but also outlines the key challenges and emerging opportunities in this rapidly evolving field.

We hope this work serves as a valuable reference for both newcomers and experienced researchers seeking to understand and apply Raman thermometry in advanced thermal characterization at the micro/nanoscale.

Oxford, OH, USA Yanan Yue

Acknowledgements

The completion of this book would not have been possible without the generous support, collaboration, and encouragement of many individuals and institutions. I would like to express my deep gratitude to the research programmes that have supported this work, as well as to the collaborating laboratories whose facilities and expertise greatly contributed to the success of the research.

I am sincerely thankful to all my colleagues and students who have contributed their time, effort, and insights during the development of this book. My appreciation goes to Dr. Shen Xu, Dr. Xiaona Huang, Dr. Dezhao Huang, and Dr. Nan Zhang, for their valuable assistance in proofreading, refining figures, and improving the overall presentation of the manuscript. I am also deeply grateful to Hongkai Zhang, Dongsheng Li, Jie Lu, Jiajie Lei, and Zechang Xiong for their dedicated work and thoughtful feedback at various stages of this project. In addition, I would like to acknowledge the efforts of former students whose earlier research laid an essential foundation for several chapters of this book.

Most importantly, I owe my deepest gratitude to my Ph.D. advisor, Prof. Xinwei Wang at Iowa State University, for introducing me to this fascinating field and for his continuous mentorship, inspiration, and encouragement throughout my academic journey. His vision and guidance have profoundly shaped my research philosophy and sustained my enthusiasm for scientific discovery.

Finally, I wish to thank my family for their unwavering love, patience, and support throughout the years. Their understanding and encouragement have been the most important source of strength that made this endeavour possible.

Competing Interests The author has no competing interests to declare that are relevant to the content of this manuscript.

Contents

Chapter 1
Thermophysical Property at Micro/ Nanoscale

Abstract This chapter provides a comprehensive overview of thermophysical properties and heat transfer phenomena at the micro- and nanoscale, which form the fundamental basis for Raman thermometry and related thermal characterization techniques. Section 1.1 introduces the background of thermal design and microscale heat transfer, emphasizing the deviation of heat conduction behavior from classical Fourier's law as the characteristic length approaches the phonon mean free path. Section 1.2 discusses the physics of thermal transport in bulk and nanoscale materials, highlighting phonon scattering, boundary effects, and size-dependent thermal conductivity. Section 1.3 reviews electro-thermal characterization methods, including the 3ω method, transient electrothermal (TET), and thermoreflectance techniques, used to probe thermal conductivity. Section 1.4 focuses on opto-thermal methods, as a non-contact and high-spatial-resolution approach for temperature measurement.

Keywords Microscale heat transfer · Thermal transport physics · Electro-thermal characterization · Opto-thermal methods

1.1 Thermal Design and Microscale Heat Transfer

Advances in semiconductor technology drive microelectronic and optoelectronic devices toward smaller size, higher operating speeds, and enhanced integration. Despite these significant advancements, thermal management becomes challenging [1, 2]. Higher device density creates localized hotspots that would undermine reliability and reduce lifetime [3]. Conventional methods, such as using high thermal conductivity metals or optimizing heat sinks, yield only marginal gains [4]. At the nanoscale, classical thermal transport models become inadequate [5], as interface scattering [6], ballistic conduction [7], and quantum effects [8] become dominant. These nanoscale mechanisms raise thermal resistance and higher temperature beyond classical predictions, and call for new understanding grounded in nanoscale thermal physics [9].

Y. Yue, *Raman Thermometry*, https://doi.org/10.1007/978-981-95-5985 5_1

Early studies relied on classical continuum theory, and Fourier's law offered an adequate description of heat flow. By the 1980s, transistor dimensions reached the micrometer scale and thermal management became a limiting factor in integrated circuits [10]. Experiments revealed local overheating that classical diffusion could not explain [4]. As device features approached the mean free paths of phonons and electrons, transport shifted from the diffusive to the ballistic regime, with few scattering events and nonlocal effects [11, 12]. Majumdar and colleagues at UC Berkeley used scanning thermal microscopy to measure temperatures at the nanoscale and showed that when the phonon mean free path is comparable to a characteristic length, conduction departs from Fourier's law [13]. Chen and colleagues at MIT demonstrated that nanostructured porous silicon can impede phonon transport effectively [14]. The same team reported near field photon tunneling between closely spaced surfaces that yields heat flux well above the blackbody limit [15]. These advances opened new approaches to thermal management and energy conversion at small scales. Nanoscale thermal management has been identified as a foundation for future electronics [16]. And its influence extended across applications such as high-performance computing and photonic devices [17–20]. As nanoscale thermal physics, materials engineering, and high-resolution measurement continue to converge, they will provide the foundation for addressing thermal-management challenges in electronic devices [9, 21] and enabling advances across energy conversion, quantum technologies, and related fields.

1.2 Physics of Thermal Transport for Bulk and Nanoscale Materials

Heat transfer can be classified into macroscopic and microscopic regimes by the ratio of a characteristic length L to the mean free path Λ of energy carriers. In the macroscopic regime, the continuum hypothesis holds, and constitutive relations such as Fourier's law can well describe temperature fields and heat flux in bulk systems. By contrast, the microscopic regime requires a carrier-based description in kinetic theory, statistical mechanics, and quantum mechanics. It describes the behavior of phonons and electrons and clarifies the mechanisms that govern energy transport.

At macroscale, heat conduction is treated as a continuous energy transfer process resulting from microscopic energy exchanges between particles within a material. Experimental observations have demonstrated that the heat flux density, defined as the thermal energy transferred per unit area per unit time, is directly proportional to the temperature gradient. This proportionality is formalized in Fourier's Law of Heat Conduction, a fundamental empirical law governing conductive heat transfer, as follows:

$$q = -k\nabla T \tag{1.1}$$

where q represents the heat flux, k is the thermal conductivity, a material property that varies with temperature. ∇T is the temperature gradient, representing the spatial rate of change of temperature. As presented in Fig. 1.1, the temperature gradient determines both the magnitude and the direction of heat flow, which always points from regions of higher temperature to regions of lower temperature.

Fourier's law assumes a continuum medium, in which materials are treated as homogeneous and continuous. However, at the atomic or molecular scale, heat conduction is governed by discrete energy carriers such as phonons and electrons. Energy transport departs from uniform diffusion and exhibits quantum behavior. When system dimensions approach or fall below the phonon mean free path, heat transport shifts from a diffusive to a ballistic regime (Fig. 1.2), rendering traditional concepts of local temperature gradients inadequate for describing energy transfer.

At nanoscale, heat transfer phenomena can no longer be accurately described using macroscopic models. Examples include one-dimensional nanowires or nanotubes and atomically thin two-dimensional materials [11], where the thermal conductivity of nanoscale materials differs significantly from that of bulk counterparts. In 2008, Chang et al. [23] experimentally confirmed the failure of Fourier's law in carbon nanotubes (CNT) and boron nitride nanotubes (BNNT) using the suspended thermal bridge method. By adjusting the length of nanotubes through platinum

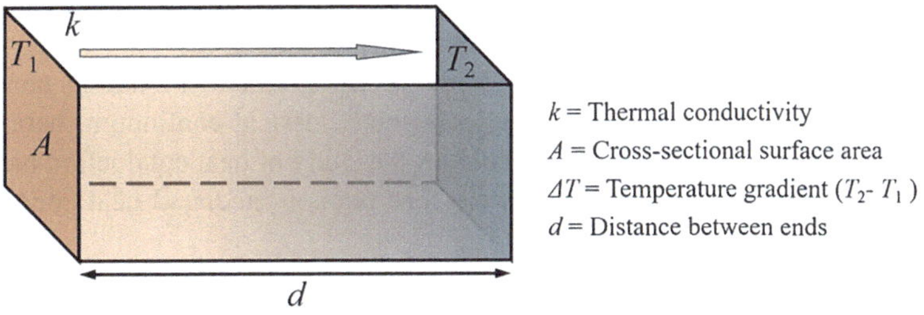

Fig. 1.1 Fourier's Law of Heat Conduction

Fig. 1.2 Sketch of a nanoscale heat transfer experimental setup [22]

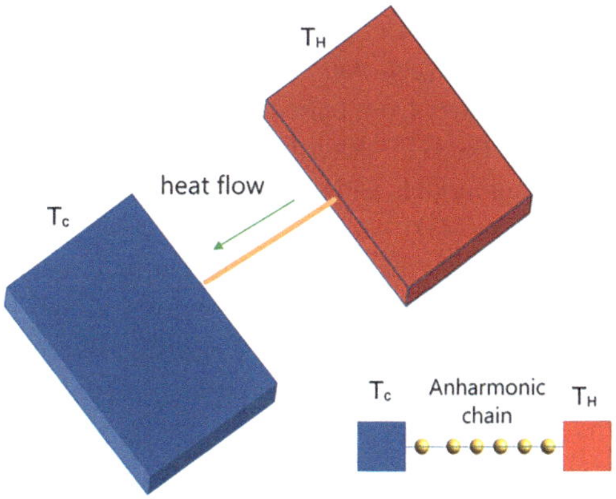

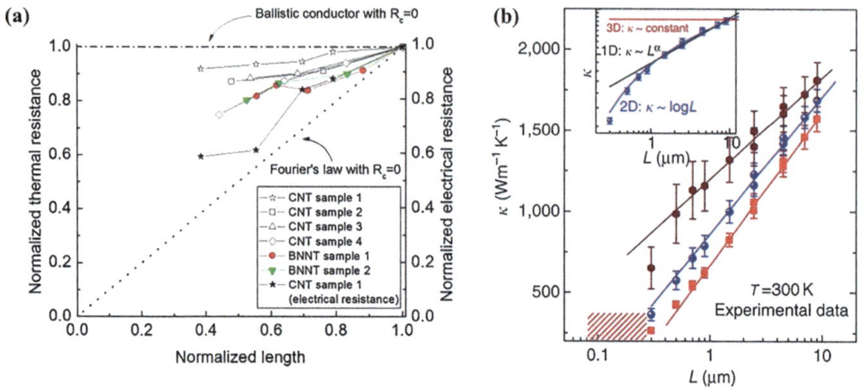

Fig. 1.3 (a) Normalized resistance vs normalized sample length for different CNT and BNNT samples. (b) Experimental results on length-dependent thermal conductivity of suspended graphene (Reprinted from ref. [23, 24], copyright (2008 and 2014), with permission from Springer Nature)

metal deposition, it was found that the thermal conductivity of both CNT and BNNT exhibited a divergence phenomenon with length. Xu et al. [24] tested the thermal conductivity of suspended graphene, and the results showed that the thermal conductivity of graphene also diverges with length (Fig. 1.3).

As anomalous heat transport draws increasing attention, modeling nanoscale phenomena demands approaches that go beyond classical continuum theory. The Boltzmann transport equation (BTE) enables the study of heat conduction based on the statistical behavior of energy carriers. For phonon-mediated heat conduction, the BTE can be expressed as:

$$\frac{\partial n_\lambda(t,\mathrm{r},\mathrm{k})}{\partial t} + \mathrm{v}_g^\lambda \cdot \nabla T \frac{\partial n_\lambda(t,\mathrm{r},\mathrm{k})}{\partial T} = \left.\frac{\partial n_\lambda(t,\mathrm{r},\mathrm{k})}{\partial t}\right|_{\mathrm{col}} \tag{1.2}$$

In Eq. 1.2, $\partial n_\lambda(t,\mathrm{r},\mathrm{k})/\partial t|_{\mathrm{col}}$ is the collision term accounts for phonon generation, annihilation, and scattering. The BTE explicitly accounts for scattering, excitation, and boundary interactions, providing a unified approach to describe thermal transport across regimes from ballistic to diffusive.

While the BTE describes mesoscale phonon kinetics, it typically uses simplified scattering models and empirical parameters. For atomic-level resolution, molecular dynamics (MD) simulations offer a powerful approach by numerically solving Newton's equations of motion for atomic systems. Through atomic level simulation modeling (Fig. 1.4), MD can directly model heat transfer mechanisms, capturing phenomena such as phonon scattering, interfacial thermal resistance, and the influence of defects in nanostructures. Statistical analysis of atomic trajectories allows the calculation of heat fluxes and temperature distributions, providing detailed insights into microscopic thermal transport processes.

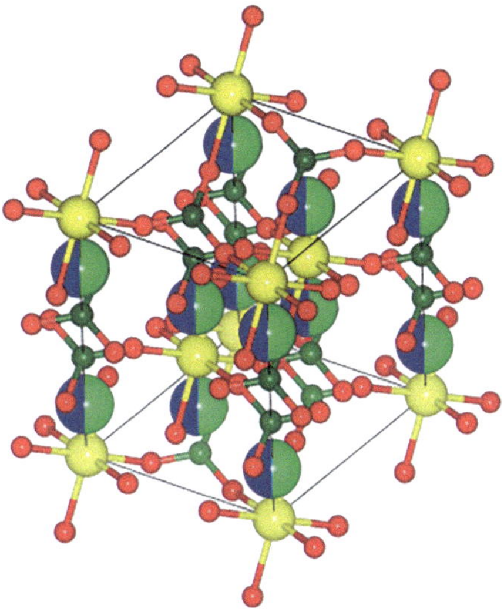

Fig. 1.4. A schematic diagram of a molecular dynamics model

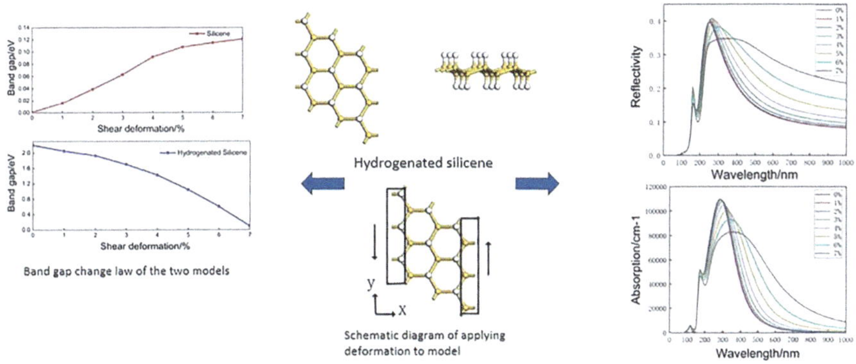

Fig. 1.5. A schematic diagram of a first-principles calculation model (Reprinted from ref. [26], with permission from Elsevier)

The accuracy of MD simulations is directly influenced by the interatomic potential, which is usually obtained from first-principles calculations (Fig. 1.5). First-principles methods are based on quantum mechanics and employ electronic structure theories, such as density functional theory, to determine material properties directly [25]. These calculations do not rely on experimental data or empirical parameters, making them suitable for studying intrinsic material properties, including crystal structures, electronic band structures, electron density distributions, and vibrational frequencies. However, they typically involve high computational costs, requiring careful selection in practical applications.

Despite the progress in theoretical and computational techniques, accurate modeling of real conditions, including complex imperfections and dynamic environments, remains challenging. Experimental works are therefore critical for validating models and revealing emergent thermal phenomena at the nanoscale [27]. However, nanoscale thermal measurements are inherently challenging. It requires precise fabrication and positioning, along with novel approaches to address issues such as thermal contact resistance, confinement effects, and interfacial influences. Advances in micro/nanofabrication have enabled the development of specialized electrical and optical measurement techniques, driving progress in high-resolution thermal characterization [27, 28]. These innovations continue to advance nanoscale thermal management, allowing increasingly precise analysis of complex thermal processes.

1.3 Electro-Thermal Characterization Methods

Electrical-based thermal characterization techniques have become indispensable for microscale and nanoscale systems, where traditional methods face inherent limitations. These techniques rely on the intrinsic temperature dependence of electrical resistance and demand precise micro/nanofabrication for implementation. Among the most widely used methods are the 3ω method [29, 30], the suspended microdevice method [31] and the T-type method [32, 33]. Each of these methods has distinct advantages and is suited to specific applications.

The 3ω method is firstly introduced by Cahill and colleagues in 1990 to measure the thermal conductivity of bulk SiO_2, and has been extended to thin films, one-dimensional (1D) and two-dimensional (2D) materials (e.g., carbon nanotubes, graphene), as well as liquids and gases. As illustrated in Fig. 1.6, a metal line is deposited on the sample surface, serving as a heater and a temperature sensor.

An alternating current with angular frequency ω is applied. It produces Joule heating at frequency 2ω and induces temperature changes at the same frequency. These temperature oscillations modulate the resistance of the metal line, leading to

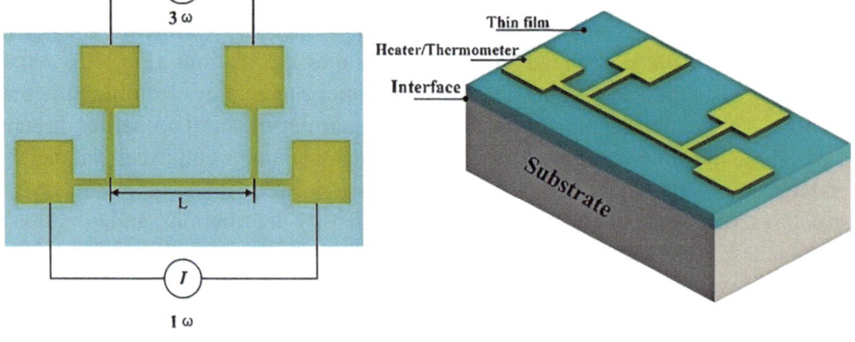

Fig. 1.6 Top and axonometric views of the 3ω measurement [29]

a third-harmonic voltage signal (3ω). The voltage is directly related to the temperature variation, allowing the extraction of thermal properties, such as thermal conductivity. For conductive samples, an insulating layer (e.g., SiO_2 or SiN) must be deposited to prevent electrical interference.

A key technical challenge of the 3ω method is the weak amplitude of the 3ω voltage signal, which is typically three orders of magnitude smaller than the applied voltage. To overcome this, the 1ω signal must be filtered out, resulting in a low signal-to-noise ratio and high sensitivity to environmental interference. The method is also limited to cases where the thermal penetration depth exceeds the width of the metal line, making it unsuitable for measuring the thermal properties of nanoscale films.

The *T*-type measurement method developed by Zhang et al. [32, 33] is an effective technique for measuring the thermal conductivity of single fibers, nanowires and nanotubes. As shown in Fig. 1.7, the method employs a *T*-shaped structure in which the test fiber is attached as a pin fin to a suspended Pt hot film positioned between two heat sinks. The Pt film serves as both heater and thermometer. A constant direct current generates uniform heat flux and the two ends are kept at ambient temperature by the heat sinks.

This method is reported with an high accuracy in measuring the thermal conductivity of a single carbon fiber [33]. In recent years, this method has also been adapted to measure the thermal boundary resistance between nanotubes and substrates such as Pt, Au, and SiO_2 [34, 35]. However, the *T*-type method also has limitations: e.g. thermal contact resistance at junctions introduces uncertainty, and the assumption of one-dimensional heat conduction may not apply to complex or anisotropic materials. Moreover, factors such as sample alignment, damage caused by sample

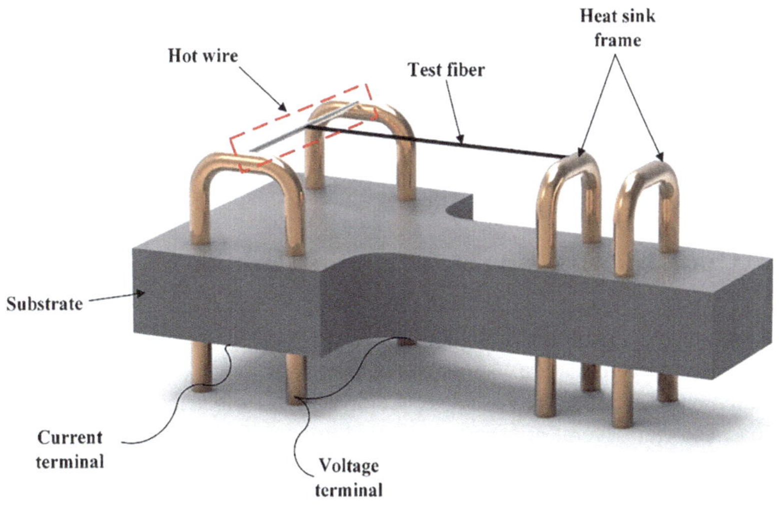

Fig. 1.7 Schematic of the experimental setup for the *T*-type method (Reprinted from ref. [27], with permission from Elsevier)

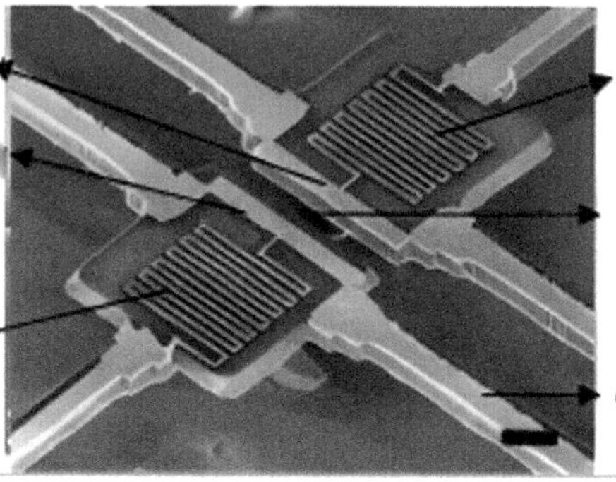

Fig. 1.8 Schematic diagram of the structure of the suspended device [31]

preparation process, and sensitivity to experimental conditions can further compromise measurement accuracy.

The suspended microdevice method, originally developed by Kim et al. at the UC Berkeley [31], was initially employed to measure the thermal conductivity of a single multi-walled carbon nanotube. Subsequent improvements, such as the incorporation of a four-probe technique, enhanced sensitivity from 10^{-7} W/mK to 10^{-9} W/mK. Additionally, structural optimizations introduced by Zhang et al. further improved the resolution to an impressive 2.5×10^{-13} W/mK [11].

This method is particularly suitable for investigating 1D and 2D materials, such as nanowires, polymer fibers, and graphene. As shown in Fig. 1.8, the experimental setup consists of suspended silicon nitride membranes (acting as heat source and sink), cantilever arms, electrodes, and temperature sensors. Applying an electric current to the heat source establishes a temperature gradient across the sample. Thermal conductivity is then determined from the measured temperature distribution and heat flow. Despite its advantages, the suspended microdevice method is limited by contact thermal resistance, especially at interfaces between high-conductivity materials (e.g., CNTs) and electrodes. This resistance can significantly reduce measurement accuracy.

1.4 Opto-Thermal Characterization Methods

Rapid advances in photonics and thermal science have led to the emergence of optical thermal characterization techniques as powerful tools for non-contact thermal measurements at micro/nanoscales. Compared to electrical methods, optical approaches offer several advantages: they eliminate the need for physical contact [1], avoiding unexpected thermal contact resistance [36], electrical leakage, and Joule heating interference associated with wire-bonding or probe-based electrical heating. Furthermore, optical techniques enable operation under extreme conditions and provide superior spatial resolution [37]. These methods leverage the

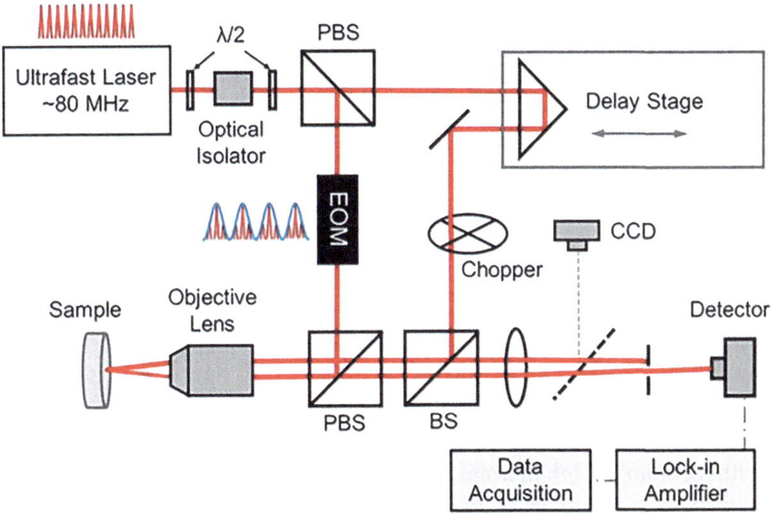

Fig. 1.9 Schematic diagram of the time-domain thermal reflectometry measurement system (Reprinted from ref. [41], with permission from its original publisher)

temperature dependence of optical signals [38], including reflectivity, scattering, and emission, each with distinct advantages and suited to specific applications.

The time-domain thermoreflectance (TDTR) is introduced as a pump-probe method employing femtosecond lasers [39], has quickly become a widely adopted technique for measuring thermal conductivity and interfacial resistance in nanoscale thin films. As shown in Fig. 1.9, this method involves splitting a pulsed laser into pump and probe beams. The pump beam heats the sample, while the probe beam detects changes in surface reflectivity caused by the temperature rise. Adjusting the optical path difference between the two beams generates a temperature-time profile. This profile is then fitted with multilayer heat transfer models to extract key thermal parameters [40].

TDTR can measure both cross-plane and in-plane thermal conductivities by adjusting the laser spot shape and size, enabling studies of anisotropic heat transport. It also provides insights into phonon contributions with different mean free paths to the total thermal conductivity. However, accuracy decreases for extremely thin samples because heat rapidly diffuses into the substrate, making it difficult to separate substrate properties from interfacial resistance.

The laser flash method is a classical and widely used non-contact technique for measuring thermal diffusivity [42]. As shown in Fig. 1.10, a short laser pulse is applied to the front surface of a sample, and the temperature rise on the rear surface is recorded. The thermal diffusivity, denoted as α, is calculated using:

$$\alpha = \frac{1.388 \times L^2}{\pi^2 \cdot t_{1/2}} \tag{1.3}$$

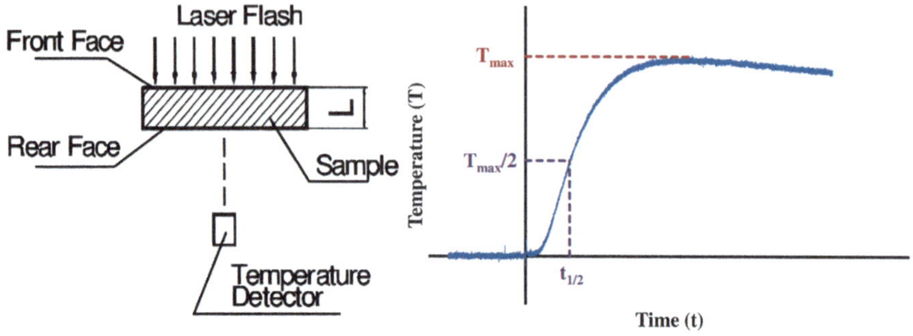

Fig. 1.10 Schematic diagram of the laser flash method [43]

where L represents the thickness of the sample, $t_{1/2}$ is the time for the rear surface to reach half of its maximum temperature rise. This technique is well-suited for materials with medium to high thermal conductivity and can also be applied to estimate interfacial resistance between films and substrates. The laser flash method applies to a broad range of materials and temperatures, making it highly flexible. However, its accuracy decreases for low-conductivity samples or when the sample thickness is extremely small.

Spectroscopic techniques for thermal characterization utilize temperature-dependent spectral features and provide semi-contact or non-contact measurement capabilities, making them feasible tools for various applications. Representative approaches include fluorescence thermometry [44], absorption/transmission spectroscopy [45], and Raman thermometry [46], each relying on distinct mechanisms and serving specific applications in nanoscale thermal analysis.

As a noncontact method, fluorescence thermometry relies on temperature-dependent fluorescence parameters such as intensity, lifetime, and spectral shape [44]. As shown in Fig. 1.11, the probe is excited by a laser, and the resulting fluorescence is analyzed to determine the temperature. Thermal conductivity is subsequently obtained through modeling. Fluorescence thermometry can be applied to microscale devices and 1D nanostructures. Probes, including organic fluorophores, quantum dots, or lanthanide complexes, can be used as the sensor to excite signals and deposited on the sample if the sample itself cannot excite fluorescence. Although this method is flexible, its sensitivity is often constrained by complex calibration procedures, instability of fluorescent emissions, and interference from ambient signals.

Absorption/transmission spectroscopy thermometry relies on temperature-dependent changes in absorption or transmission spectra. Back in 1979, Weakliem et al. [49] measured the absorption spectrum of silicon between 297 and 473 K and reported a redshift with increasing temperature. Spectral parameters such as intensity, peak position, and full width at half maximum (FWHM) serve as temperature indicators. A typical setup (Fig. 1.12) consists of a light source, detector, signal processor, and sample. By comparing recorded absorption spectra with calibrated reference data, the sample temperature can be determined. However, the accuracy of

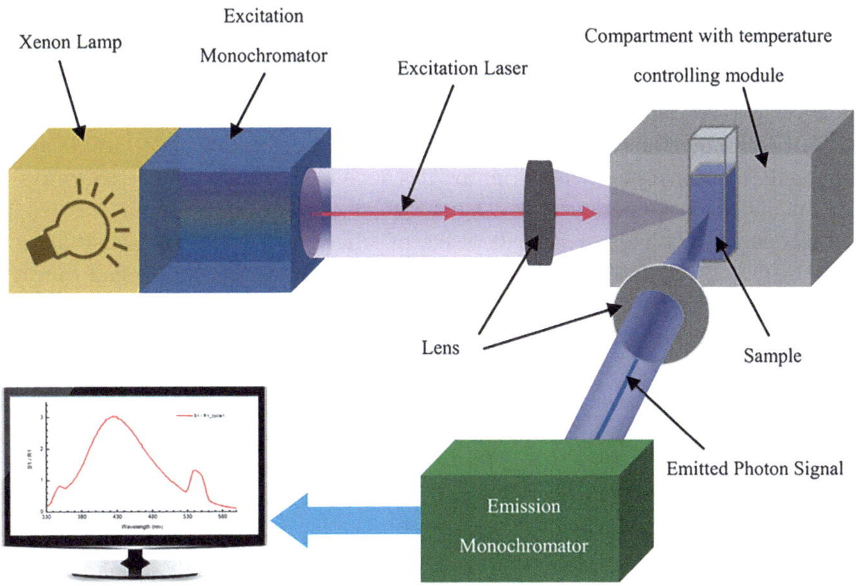

Fig. 1.11 Schematic diagram of fluorescence temperature measurement [47]

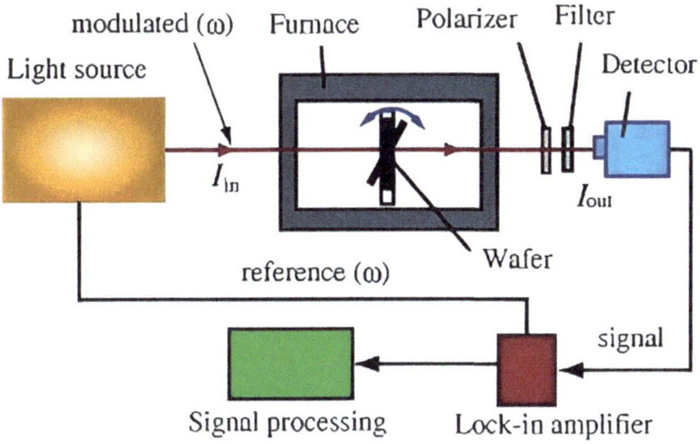

Fig. 1.12 Schematic diagram of absorption spectroscopy thermometry (Reprinted from ref. [48], with permission from original publisher)

absorption-based thermometry depends strongly on detailed knowledge of the material's optical properties. Broad spectral linewidths and difficulties in resolving peak positions often limit precision. Furthermore, the setup is complex, and calibration is time-consuming.

Raman thermometry offers several advantages over conventional thermometry techniques [46]. It operates non-invasively, eliminating measurement artifacts caused by other optical method. Moreover, Raman thermometry uniquely combines temperature measurement with material structural analysis through unique vibrational patterns. These attributes make it particularly suitable for analyzing thermal properties in micro/nanoscale materials, including 2D materials, semiconductor heterostructures, and nanoscale devices.

The technique relies on the temperature dependence of Raman spectral features, enabling direct measurements of properties such as thermal conductivity, diffusivity, and localized temperatures in 2D materials. In 1970, Hart et al. demonstrated the temperature sensitivity of Raman peaks in silicon (e.g., intensity, position, and FWHM). And the method has evolved to include both steady-state and transient modes [50]. Using a laser to excite Raman scattering in the sample (Fig. 1.13), the scattered light is analyzed to determine temperature through pre-calibrated spectral correlations. Coupled with physical modeling, this technique also enables the extraction of thermal transport properties.

Raman thermometry offers distinct advantages over conventional methods, providing non-contact temperature measurement and thermal analysis at micro/nanoscales. As temperature variations in microdevices are becoming increasingly concentrated in small areas, making conventional thermal measurement less effective. In contrast, the excitation laser in a Raman system can be focused to submicron spots (~500 nm under a 100× objective), enabling highly localized temperature probing. A further key advantage of Raman-based thermometry is its material specificity: it not only measures temperature but also identifies the material being probed through its characteristic Raman spectrum.

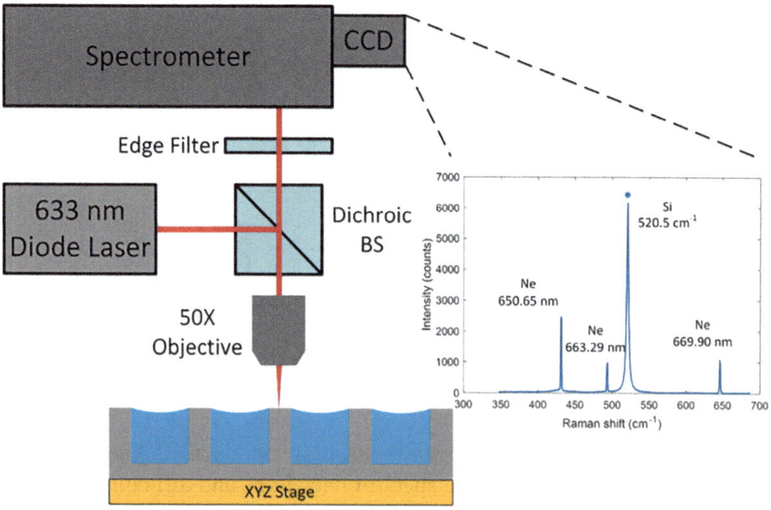

Fig. 1.13 Schematic diagram of Raman spectroscopy thermometry

References

1. Cahill, D.G., Goodson, K., Majumdar, A.: Thermometry and thermal transport in micro/nanoscale solid-state devices and structures. J. Heat Transfer. **124**(2), 223–241 (2001)
2. Liu, Y., Duan, X., Shin, H.-J., Park, S., Huang, Y., Duan, X.: Promises and prospects of two-dimensional transistors. Nature. **591**(7848), 43–53 (2021)
3. Pop, E.: Energy dissipation and transport in nanoscale devices. Nano Res. **3**(3), 147–169 (2010)
4. Chen, G.: Introduction. In: Nanoscale energy transport and conversion: a parallel treatment of electrons, molecules, phonons, and photons, pp. 3–42. Oxford University Press (2005)
5. Tien, C.L., Chen, G.: Challenges in microscale conductive and radiative heat transfer. J. Heat Transfer. **116**(4), 799–807 (1994)
6. Volz, S.G., Chen, G.: Molecular dynamics simulation of thermal conductivity of silicon nanowires. Appl. Phys. Lett. **75**(14), 2056–2058 (1999)
7. Minnich, A.J., Johnson, J.A., Schmidt, A.J., Esfarjani, K., Dresselhaus, M.S., Nelson, K.A., Chen, G.: Thermal conductivity spectroscopy technique to measure phonon mean free paths. Phys. Rev. Lett. **107**(9), 095901 (2011)
8. Rego, L.G.C., Kirczenow, G.: Quantized thermal conductance of dielectric quantum wires. Phys. Rev. Lett. **81**(1), 232–235 (1998)
9. Maldovan, M.: Phonon wave interference and thermal bandgap materials. Nat. Mater. **14**(7), 667–674 (2015)
10. Majumdar, A.: Scanning thermal microscopy. Annu. Rev. Mater. Res. **29**, 505–585 (1999)
11. Zhang, Z., Ouyang, Y., Cheng, Y., Chen, J., Li, N., Zhang, G.: Size-dependent phononic thermal transport in low-dimensional nanomaterials. Phys. Rep. **860**, 1–26 (2020)
12. Hopkins, P.E., Norris, P.M.: Effects of joint vibrational states on thermal boundary conductance. Nanoscale Microscale Thermophys. Eng. **11**(3–4), 247–257 (2007)
13. Majumdar, A.: Microscale heat conduction in dielectric thin films. J. Heat Transfer. **115**(1), 7–16 (1993)
14. Song, D., Chen, G.: Thermal conductivity of periodic microporous silicon films. Appl. Phys. Lett. **84**(5), 687–689 (2004)
15. Hu, L., Narayanaswamy, A., Chen, X., Chen, G.: Near-field thermal radiation between two closely spaced glass plates exceeding Planck's blackbody radiation law. Appl. Phys. Lett. **92**(13), 133106 (2008)
16. Song, B., Thompson, D., Fiorino, A., Ganjeh, Y., Reddy, P., Meyhofer, E.: Radiative heat conductances between dielectric and metallic parallel plates with nanoscale gaps. Nat. Nanotechnol. **11**(6), 509–514 (2016)
17. Kim, S.E., Mujid, F., Rai, A., Eriksson, F., Suh, J., Poddar, P., Ray, A., Park, C., Fransson, E., Zhong, Y., Muller, D.A., Erhart, P., Cahill, D.G., Park, J.: Extremely anisotropic van der Waals thermal conductors. Nature. **597**, 660–665 (2021)
18. Stamper, C., Cortie, D., Nazrul-Islam, S.M.K., Rahman, M.R., Yu, D., Yang, G., Al-Mamun, A., Wang, X., Yue, Z.: Phonon engineering in thermal materials with nano-carbon dopants. Appl. Phys. Rev. **11**(2), 021336 (2024)
19. Warzoha, R.J., Wilson, A.A., Donovan, B.F., Donmezer, N., Giri, A., Hopkins, P.E., Choi, S., Pahinkar, D., Shi, J., Graham, S., Tian, Z., Ruppalt, L.: Applications and impacts of nanoscale thermal transport in electronics packaging. J. Electron. Packag. **143**(2), 020804 (2021)
20. Hu, Y., Xu, J., Ruan, X., Bao, H.: Defect scattering can lead to enhanced phonon transport at nanoscale. Nat. Commun. **15**(1), 3304 (2024)
21. Li, Y., Li, W., Han, T., Zheng, X., Li, J., Li, B., Fan, S., Qiu, C.-W.: Transforming heat transfer with thermal metamaterials and devices. Nat. Rev. Mater. **6**(6), 488–507 (2021)

22. Benenti, G., Donadio, D., Lepri, S., Livi, R.: Non-Fourier heat transport in nanosystems. Riv. Nuovo Cimento. **46**(3), 105–161 (2023)
23. Chang, C.W., Okawa, D., Garcia, H., Majumdar, A., Zettl, A.: Breakdown of Fourier's law in nanotube thermal conductors. Phys. Rev. Lett. **101**(7), 075903 (2008)
24. Xu, X., Pereira, L.F.C., Wang, Y., Wu, J., Zhang, K., Zhao, X., Bae, S., Tinh Bui, C., Xie, R., Thong, J.T.L., Hong, B.H., Loh, K.P., Donadio, D., Li, B., Özyilmaz, B.: Length-dependent thermal conductivity in suspended single-layer graphene. Nat. Commun. **5**(1), 3689 (2014)
25. Martin, R.M.: Density Functional Theory: Foundations. In: Electronic Structure, pp. 119–134. Cambridge University Press, Cambridge (2004)
26. Yao, Y., Liu, G.: First-principles calculation to investigate the influence of shear deformation on the electronic structure and optical properties of hydrogenated silicene. Comput. Theor. Chem. **1207**, 113506 (2022)
27. Xian, Y., Zhang, P., Zhai, S., Yuan, P., Yang, D.: Experimental characterization methods for thermal contact resistance: a review. Appl. Therm. Eng. **130**, 1530–1548 (2018)
28. Wang, J., Wang, Z., Yang, K., Chen, N., Ni, J., Song, J., Li, Q., Sun, F., Liu, Y., Fan, T.: Enhanced heat transport capability across boron nitride/copper interface through inelastic phonon scattering. Adv. Funct. Mater. **32**(40), 2206545 (2022)
29. Cahill, D.G.: Thermal conductivity measurement from 30 to 750 K: the 3ω method. Rev. Sci. Instrum. **61**(2), 802–808 (1990)
30. Lu, L., Yi, W., Zhang, D.L.: 3ω method for specific heat and thermal conductivity measurements. Rev. Sci. Instrum. **72**(7), 2996–3003 (2001)
31. Kim, P., Shi, L., Majumdar, A., Mceuen, P.L.: Thermal transport measurements of individual multiwalled nanotubes. Phys. Rev. Lett. **87**(21), 215502 (2001)
32. Fujii, M., Zhang, X., Xie, H., Ago, H., Takahashi, K., Ikuta, T., Abe, H., Shimizu, T.: Measuring the thermal conductivity of a single carbon nanotube. Phys. Rev. Lett. **95**(6), 065502 (2005)
33. Zhang, X., Fujiwara, S., Fujii, M.: Measurements of thermal conductivity and electrical conductivity of a single carbon fiber. Int. J. Thermophys. **21**(4), 965–980 (2000)
34. Hirotani, J., Ikuta, T., Nishiyama, T., Takahashi, K.: Thermal boundary resistance between the end of an individual carbon nanotube and a au surface. Nanotechnology. **22**(31), 315702 (2011)
35. Hirotani, J., Ikuta, T., Nishiyama, T., Takahashi, K.: Measuring the thermal boundary resistance of van der Waals contacts using an individual carbon nanotube. J. Phys. Condens. Matter. **25**(2), 025301 (2013)
36. Shen, S., Henry, A., Tong, J., Zheng, R., Chen, G.: Polyethylene nanofibres with very high thermal conductivities. Nat. Nanotechnol. **5**(4), 251–255 (2010)
37. Chakarvarti, S.K., Gehlawat, D., Manocha, A.: Thermal interface materials for thermal management of microelectronic devices: a review. J. Therm. Anal. Calorim. **150**(12), 8847–8860 (2025)
38. Palik, E.D.: Chapter 1 - Introductory remarks. In: Handbook of Optical Constants of Solids, pp. 3–9. Academic, Boston (1985)
39. Cahill, D.G.: Analysis of heat flow in layered structures for time-domain thermoreflectance. Rev. Sci. Instrum. **75**(12), 5119–5122 (2004)
40. Feldman, A.: Algorithm for solutions of the thermal diffusion equation in a stratified medium with a modulated heating source. High Temp.-High Pressures. **31**(3), 293–298 (1999)
41. Jiang, P., Qian, X., Yang, R.: Tutorial: time-domain thermoreflectance (TDTR) for thermal property characterization of bulk and thin film materials. J. Appl. Phys. **124**(16), 161103 (2018)
42. Parker, W.J., Jenkins, R.J., Butler, C.P., Abbott, G.L.: Flash method of determining thermal diffusivity, heat capacity, and thermal conductivity. J. Appl. Phys. **32**(9), 1679–1684 (1961)
43. Blumm, J.: The laser flash technique: a widespread technology for measurement of the thermal diffusivity of solids and liquids. Int. J. Thermophys. **46**(3), 39 (2025)
44. Brites, C.D.S., Marin, R., Suta, M., Carneiro Neto, A.N., Ximendes, E., Jaque, D., Carlos, L.D.: Spotlight on luminescence thermometry: basics, challenges, and cutting-edge applications. Adv. Mater. **35**(36), 2302749 (2023)

45. Ozaki, Y., Šašić, S.: Introduction to Raman spectroscopy. In: Pharmaceutical Applications of Raman Spectroscopy, pp. 1–28 (2007)
46. Balandin, A.A., Ghosh, S., Bao, W., Calizo, I., Teweldebrhan, D., Miao, F., Lau, C.N.: Superior thermal conductivity of single-layer graphene. Nano Lett. **8**(3), 902–907 (2008)
47. Okabe, K., Sakaguchi, R., Shi, B., Kiyonaka, S.: Intracellular thermometry with fluorescent sensors for thermal biology. Pflugers Arch. **470**(5), 717–731 (2018)
48. Toyoda, Y., Seo, T., Iuchi, T.: Non-contact temperature measurement of silicon wafers based on the combined use of transmittance and radiance. Measurement. **51**, 393–399 (2014)
49. Weakliem, H., Redfield, D.: Temperature dependence of the optical properties of silicon. J. Appl. Phys. **50**(3), 1491–1493 (1979)
50. Hart, T.R., Aggarwal, L.B.L.R.: Temperature dependence of Raman scattering in silicon. Phys. Rev. B Condens. Matter. **1**(2), 638–642 (1970)

Chapter 2
Physical Principles of Raman Thermometry

Abstract This chapter introduces the fundamental physical principles underlying Raman thermometry. Section 2.1 reviews the fundamentals of Raman spectroscopy, including the theory of inelastic light scattering. Section 2.2 describes the typical Raman spectrometer setup, covering excitation sources, optical components, and detectors. Section 2.3 focuses on the temperature dependence of Raman spectra. Section 2.4 summarizes typical applications of Raman thermometry in materials science, microelectronics, and nanotechnology, demonstrating its capability to measure temperature, probe thermal transport, and evaluate interfacial thermal resistance. This chapter discusses physical principles to practical thermal characterization methodologies, forming the theoretical foundation for the implementation of Raman-based thermal metrology.

Keywords Raman spectroscopy · Inelastic light scattering · Temperature dependence · Non-contact measurement

2.1 Fundamentals of Raman Spectroscopy

Raman spectroscopy is an analytical technique that probes the vibrational and rotational modes of molecules to provide chemical and structural information about materials. Discovered by C. V. Raman et al. in 1928 [1], Raman effecst is based on the inelastic scattering phenomenon wherein incident photons exchange energy with molecular vibrations, resulting in scattered light with shifted frequency. This frequency shift, known as the Raman shift, yields a characteristic spectral signature that enables precise material identification.

When incident photons interact with a sample, the oscillating electromagnetic field induces polarization of the electron cloud, transiently exciting the molecule to a virtual energy state. This unstable intermediate state decays almost instantaneously, emitting a scattered photon. If the molecule returns to its original vibrational state, the scattered photon has the same energy as the incident photon

Y. Yue, *Raman Thermometry*, https://doi.org/10.1007/978-981-95-5985-5_2

(Rayleigh scattering). As an elastic scattering process, Rayleigh scattering accounts for the majority of scattering. However, if the molecule transitions to a different vibrational level, energy is exchanged, producing inelastically scattered photons with altered energy, which is called Raman scattering [2, 3].

As shown in Fig. 2.1, when the molecule absorbs energy during the scattering event, it transitions to a higher vibrational level, and the scattered photon loses an equivalent amount of energy. This produces a shift toward lower energy (redshift) in the emission spectrum, referred as Stokes Raman scattering. Conversely, if the molecule relaxes to a lower vibrational state, releasing energy to the photon, the emitted photon gains this energy, resulting in a shift toward higher energy (blueshift), known as anti-Stokes Raman scattering [4]. Although Stokes and anti-Stokes scattering are equally probable from a quantum mechanical perspective, the Boltzmann distribution ensures that, at equilibrium, most molecules occupy their lowest vibrational state. As a result, Stokes scattering is far more common and produces a stronger signal, making it the primary focus of Raman spectroscopy.

Figure 2.2 schematically illustrates the interpretation of Raman scattering, wherein the recorded Raman spectrum reflects the energy differences between

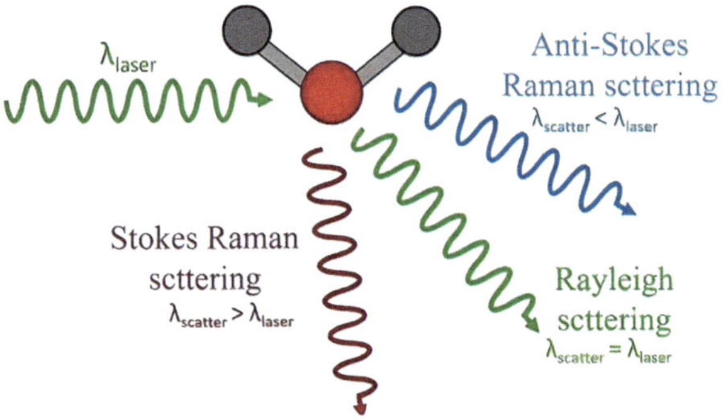

Fig. 2.1 Schematic of the Raman scattering principle

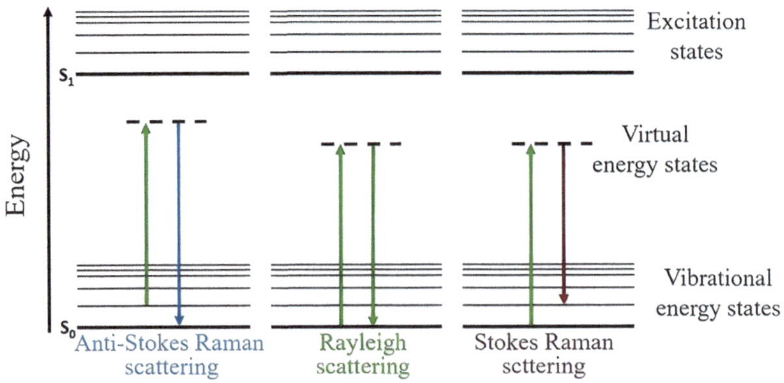

Fig. 2.2 Fundamental principle of Raman spectroscopy in quantum theoretical interpretation

vibrational energy levels of the molecule. Vibrational states correspond to specific vibrational modes of a molecule; for polyatomic molecules, multiple vibrational modes exist, each associated with unique sets of quantized energy levels. Assuming a molecule initially resides in its ground vibrational state E_0, irradiation by a laser of frequency ω excites the system by absorbing a photon of energy $h\omega$, promoting the molecule to a virtual energy state ($E_0 + h\omega$). This virtual state is transient and inherently unstable, leading the molecule to relax rapidly by emitting scattered light [3].

The absolute wavelength of scattered light depends on the excitation source wavelength. Therefore, Raman spectroscopic measurements are typically expressed as the Raman shift, the energy difference between the incident and scattered photons, regardless of the laser wavelength. The Raman shift is determined by the equation:

$$\omega = \left(\frac{1}{\lambda_0} - \frac{1}{\lambda_1} \right) \times 10^7 \left[\mathrm{cm}^{-1} \right] \tag{2.1}$$

where λ_0 is the incident laser wavelength and λ_1 is the scattered photon wavelength (both in nm). This equation ensures that Raman shifts are intrinsic to molecular energy structure and remain consistent for a given material under different experimental conditions [3].

Figure 2.3 presents representative spectra of two crystalline materials: single-crystal silicon and monolayer graphene. As shown in Fig. 2.3a, silicon exhibits a sharp first-order Raman peak at approximately 520 cm^{-1} at room temperature, corresponding to the triply degenerate optical phonon mode at the Brillouin zone center. The peak position and linewidth are highly sensitive to temperature, stress, and crystalline quality, making this feature widely used for thermal and mechanical characterization of silicon devices. Figure 2.3b depicts the Raman spectrum of monolayer graphene, characterized by three prominent features: the G peak, the D

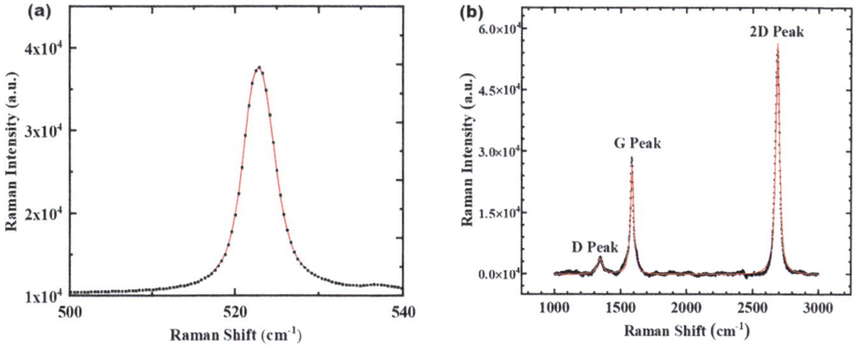

(a) Schematic Raman spectrum of Silicon

(b) Schematic Raman spectrum of Graphene

Fig. 2.3 Representative Raman spectra of common materials: (**a**) silicon, (**b**) graphene peak, and the 2D peak [5]. The G peak is centered around 1580 cm^{-1}, arises from in-plane vibrations of sp^2-hybridized carbon atoms and serves as a key signature of graphene. Its position is often used to infer the number of graphene layers and is sensitive to mechanical strain. The D peak originates from disorder-induced vibrations and off-zone-center phonons, is absent in pristine graphene and instead appears in defective samples, providing a diagnostic measure of structural imperfections. The 2D peak is a second-order band resulting from a double-resonance process, is influenced by the stacking sequence of graphene layers and depends on the excitation wavelength. Apart from these features, weaker second-order Raman bands may also appear, offering insights into electron–phonon and electron–electron interactions [6].

Raman spectroscopy is a powerful characterization approach for investigating molecular vibrational and rotational energy levels, enabling quantitative analysis of intermolecular and intramolecular interactions. Although Raman scattering is inherently weak (typically representing 10^{-6} to 10^{-12} of the incident light intensity), the advances in laser sources, high-throughput spectrometers, CCD detectors, and computational spectral analysis have transformed it to an indispensable analytical method. It is widely applied across disciplines such as physics, chemistry, biology, materials science, and microelectronics, as well as in industrial applications. Importantly, Raman spectroscopy is powerful for probing low-dimensional nanomaterials, with its capabilities now extending to temporal, spatial, and depth-resolved measurements. Specialized methods such as resonance-enhanced Raman spectroscopy, time-resolved Raman spectroscopy, micro/near-field Raman spectroscopy, surface-enhanced Raman spectroscopy, and Raman spectral imaging have further expanded its scope, enabling studies of dynamic processes, micro/nanoscale systems, and interfacial phenomena [7].

2.2 Typical Raman Spectrometer Setup

A standard Raman spectrometer system consists of several key components designed to efficiently excite the sample with a monochromatic light source and collect the scattered photons for spectral analysis. As show in Fig. 2.4, the typical setup includes:

2.2.1 Excitation Source (Laser)

The excitation source is usually a continuous-wave (CW) laser with a narrow linewidth and high spectral purity. Commonly used laser wavelengths include 488 nm, 514.5 nm, 532 nm, 633 nm, and 785 nm, chosen based on sample absorption characteristics and desired Raman signal strength. The laser is directed through optical components to focus onto the sample with a well-defined spot size, often on the order of 0.5–1 μm for micro-Raman systems.

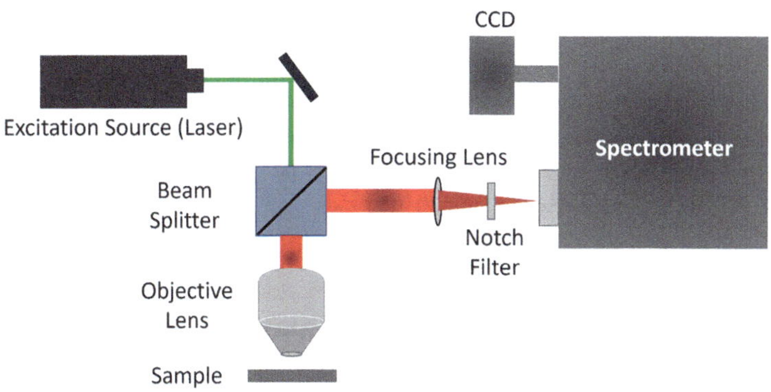

Fig. 2.4 Typical Raman spectrometer setup

2.2.2 Optical Path and Filtering

Before reaching the sample, the laser passes through beam shaping optics and a notch filter to remove some spectral signals. After interacting with the sample, the backscattered light, which contains both Rayleigh and Raman scattered photons, is collected by the same objective lens (in a backscattering geometry) and directed toward the spectrometer. Additional edge or notch filters are used to suppress the much stronger Rayleigh scattering, allowing only the Raman-shifted light to pass through.

2.2.3 Microscope and Sample Stage

A high-resolution optical microscope, integrated into the system, allows for precise focusing and spatial alignment. The microscope objective (typically 50× or 100×) serves a dual purpose: focusing the laser on the sample and collecting scattered light. The sample can be mounted on a motorized XYZ translation stage, enabling spatial mapping and site-specific measurements with submicron accuracy.

2.2.4 Spectrometer

The Raman-shifted light is dispersed by a diffraction grating inside the spectrometer, which separates different wavelengths based on their frequency shift relative to the excitation laser. The resolution of the spectrometer depends on the grating density and optical design, typically allowing spectral resolution on the order of 0.5–2 cm^{-1}.

2.2.5 CCD Detector

The dispersed Raman signal is detected using a charge-coupled device (CCD) detector, which is often cooled to reduce thermal noise and improve signal-to-noise ratio. The CCD captures the intensity of Raman peaks across the spectral range of interest, producing a full Raman spectrum.

2.3 Temperature Dependence of Raman Spectra

Raman spectroscopy provide valuable insights into the properties of materials, offering information on molecular composition, phonon lifetime through peak linewidth, and phonon population via peak intensity [8]. Beyond characterizing chemical structures, Raman scattering has been extensively applied to reveal physical properties such as temperature, stress, and grain size. Among these properties, the temperature-induced changes in Raman spectra are particularly significant and form the physical basis for Raman thermometry. The temperature dependence of Raman peak frequency arises from two primary lattice dynamic effects: thermal expansion increases interatomic spacing with rising temperature, weakening bond restoring forces and lowering vibrational frequencies, which show as a redshift in Raman peak position. More significantly, anharmonic phonon-phonon interactions become increasingly active at elevated temperatures. These anharmonic effects, dominated by phonon scattering processes, renormalize phonon energies and cause Raman peaks to shift to lower frequencies and broaden. By establishing the correlation between Raman peak position and temperature under controlled conditions, researchers can perform non-contact, spatially resolved temperature measurements with microscale or even nanoscale resolution. When combined with heat transfer models, this spectroscopic approach enables extraction thermophysical properties of materials [9].

Raman thermometry offers distinct advantages over conventional temperature measurement techniques, particularly for micro/nanoscale thermal characterization. Traditional contact-based methods, such as thermocouples, require direct physical attachment to the sample [10, 11]. As devices continue to downscale, particularly in microelectronic systems and 2D materials, conventional methods is more difficult to achieve the required spatial resolution for precise thermal probing. Raman systems address this challenge by focusing lasers to sub-micron spot sizes [12], enabling localized temperature probing. Importantly, Raman spectra not only provide temperature information specific to the measured material but also identify the material [11, 13].

Raman spectra exhibit three primary temperature-dependent features that serve as the foundation for thermometric applications: peak position (Raman shift), peak intensity, and linewidth (full width at half maximum, FWHM) [11, 14, 15]. These spectral parameters are intrinsically linked to the lattice dynamics of the material.

Specifically, the temperature-dependent property of phonon energies, populations, and lifetimes make them as reliable thermal probes. Generally, an increase in temperature causes a redshift in the Stokes Raman shift of a characteristic peak, a broadening of the linewidth, and a reduction in peak intensity. For relatively small temperature changes (~100 °C), these variations maintain a linear relationship with temperature, enabling quick and accurate temperature measurement in samples.

The Raman shift of characteristic peaks originates from the dipoles created by the chemical structure in the material's lattice. For most materials, the Raman shift of prominent peaks decreases with increasing temperature. While both Stokes and anti-Stokes signals shift simultaneously, the Stokes signal is significantly stronger and thus more reliable for temperature measurement. The relationship between the temperature and the Stokes signal shift can be expressed as follows [17].

$$\omega(T) = \omega_0 + A\left(1 + \frac{2}{e^{\frac{h\omega_0}{4\pi kT}} - 1}\right) + B\left(1 + \frac{3}{e^{\frac{h\omega_0}{4\pi kT}} - 1} + \frac{3}{\left(e^{\frac{h\omega_0}{4\pi kT}} - 1\right)^2}\right) \tag{2.2}$$

where $\omega(T)$ is the Raman frequency shift, T is the absolute temperature of the sample, h is the Planck constant, and k the Boltzmann constant. ω_0 is the Raman frequency at room temperature, and A and B are constants specific to materials. As shown in Fig. 2.5, silicon can be shown as an example where Raman peak shifts are highly sensitive to temperature variations, making it a commonly used material for Raman thermometry. However, temperature gradients can induce stress in materials, leading to additional shifts in the Raman frequency. This stress-related effect reduces the accuracy of temperature measurements based solely on Raman shift. In cases that the temperature increase is modest, the stress effects are often negligible, allowing Raman shift measurements to serve as accurate temperature probes.

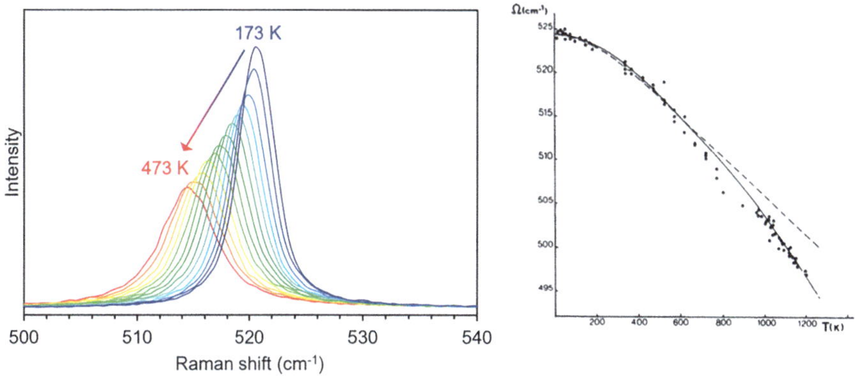

Fig. 2.5 Temperature-dependent Raman frequency shift in silicon (reprinted from ref. [16], Copyright(2023), with permission from Elsevier)

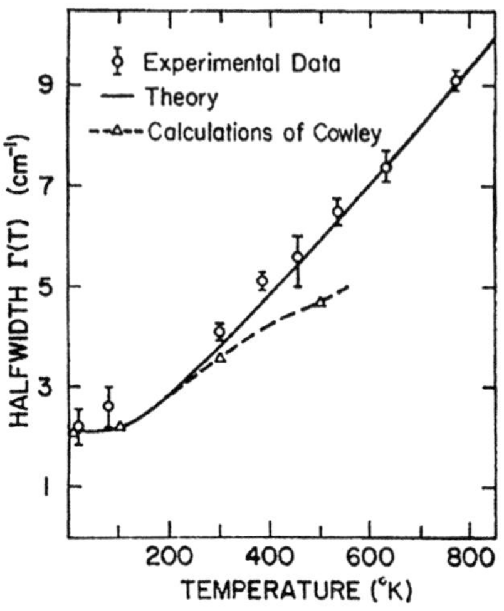

Fig. 2.6 Temperature-dependent FWHM in silicon (reprinted from ref. [16], Copyright(2023), with permission from Elsevier)

Another key intrinsic property is the Raman peak linewidth, as illustrated in Fig. 2.6. The linewidth is determined by the lifetime of optical phonons during the Raman scattering process. As temperature changes, the phonon lifetime is affected, resulting in corresponding changes in the Raman peak's linewidth. Generally, the linewidth of a Raman peak broadens with increasing temperature. The relationship between temperature and linewidth can be expressed as [17]:

$$\Gamma(T) = C\left(1 + \frac{2}{e^{\frac{h\omega_0}{4\pi kT}} - 1}\right) + D\left(1 + \frac{3}{e^{\frac{h\omega_0}{4\pi kT}} - 1} + \frac{3}{\left(e^{\frac{h\omega_0}{4\pi kT}} - 1\right)^2}\right) \tag{2.3}$$

Here, Γ is the Raman peak linewidth. Close to eq. (2.2), C and D are material-specific constants. However, in practical applications, linewidth is rarely used for temperature determination due to its lower sensitivity to temperature changes. Accurately measuring linewidth requires high-quality Raman signals, and insufficient signal quality may result in significant errors in temperature estimation. Compared to Raman shift, linewidth is less influenced by stress, making it a potential alternative indicator for temperature measurement in cases involving substantial material stress.

The intensity of Raman peaks varies with temperature due to changes in the population of vibrational states. At different temperatures, the density of atoms in excited vibrational states differs, leading to temperature-dependent changes in Raman intensity under the same incident light and acquisition settings. Notably, the Stokes and anti-Stokes signals have opposite trends in their intensity variation with temperature: the Stokes intensity decreases as temperature increases, while the

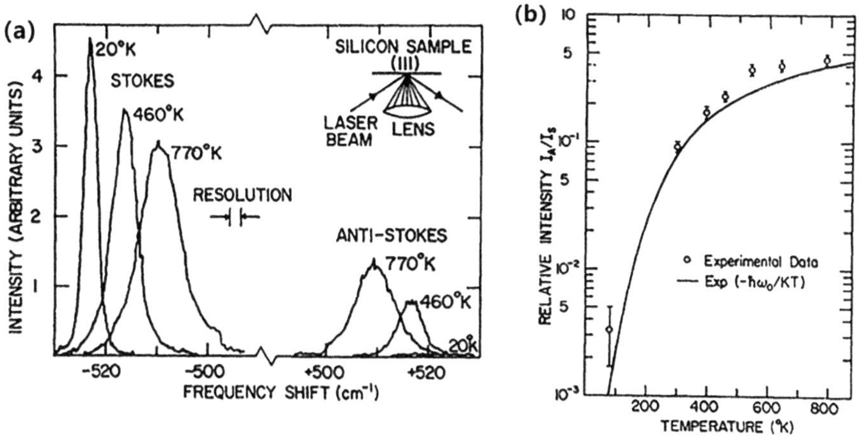

Fig. 2.7 Correlation between Raman spectral intensity and temperature in silicon (**a**) Variation of the anti-Stokes to Stokes intensity ratio with temperature;(**b**) Temperature dependence of the Stokes and anti-Stokes intensities plotted separately (reprinted from ref. [16], Copyright(2023), with permission from Elsevier)

anti-Stokes intensity increases. For silicon, the ratio of anti-Stokes to Stokes intensity can be expressed as [17]:

$$\frac{I_{\mathrm{s}}}{I_{\mathrm{as}}} = \frac{\alpha_{\mathrm{i}} + \alpha_{\mathrm{s}}}{\alpha_{\mathrm{i}} + \alpha_{\mathrm{as}}} + \left(\frac{\omega_{\mathrm{i}} - \omega_{\mathrm{R}}}{\omega_{\mathrm{i}} + \omega_{\mathrm{R}}} \right)^4 \exp\left(\frac{h\omega_{\mathrm{R}}}{2\pi kT} \right) \tag{2.4}$$

where α_{i}, α_{s}, α_{as} are the absorbance of the material at the incident light frequency, the Stokes frequency and the anti-Stokes frequency, respectively. ω_{R} is the frequency of the Raman peak; ω_i is the incident light frequency. From this equation, it is evident that the intensity ratio of anti-Stokes to Stokes signals is a direct function of temperature, as illustrated in Fig. 2.7a. This feature allows for temperature determination using Raman intensity measurements. However, as shown in Fig. 2.7b, anti-Stokes scattering is often challenging to measure with commercial Raman systems, as its frequency is higher than the incident laser frequency and may fall outside the detection range of the system. Additionally, anti-Stokes signals are much weaker than Stokes signals, which reduces the accuracy of intensity based methods for temperature measurement.

2.4 Typical Applications of Raman Thermometry

Raman thermometry has proven to be an effective tool for temperature measurement in a wide range of materials, including silicon [4, 6, 11], carbon nanotubes [13–17], graphene [18–20] . It has also been widely utilized to investigate heat transfer under various physical conditions. Its non-invasive nature and high spatial resolution make it uniquely suited for probing microscale and nanoscale thermal processes where conventional thermometry techniques are impractical. In the following context, we highlight representative applications that illustrate the capabilities of Raman-based thermal measurements.

2.4.1 Thermal Conductivity in Graphene

Graphene's remarkable thermal properties have been studied extensively using Raman. Balandin and colleagues systematically investigated temperature-dependent Raman shifts in graphene layers of varying thicknesses. Their work established distinct temperature coefficients for single-layer and multilayer graphene, providing foundational data for thermal characterization in graphene [18, 19].

Besides this work, Raman thermometry have also been leveraged as the dual functionality as both a heater and a thermometer [21]. Figure 2.8 presents an example, suspending graphene over microscopic holes allows localized laser heating at the center, inducing radial heat flow [20]. By tracking the resulting temperature gradient through Raman peak shifts, such as the 2D peak with a temperature coefficient of -0.072 cm^{-1}/K [22], graphene's in-plane thermal conductivity can be calculated. Experimental results indicate that the thermal conductivity decreases from about 1800 W/mK at 325 K to around 710 W/mK at 500 K [20], demonstrating graphene's pronounced temperature-dependent behavior. Further exploration of these mechanisms and their dependence on structural and environmental factors will be detailed in Chap. 3.

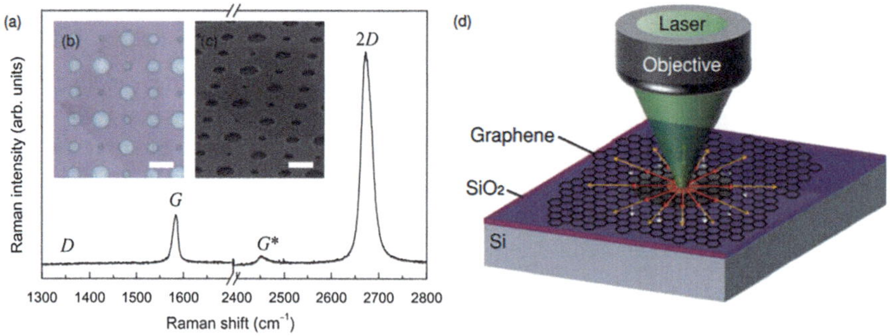

Fig. 2.8 Schematic of thermal conductivity measurement of graphene based on Raman thermometry: (**a**) Raman spectrum, (**b**) optical and (**c**) SEM image of graphene, (**d**) Schematic of measurement (reprinted from ref. [20], Copyright (2013), with permission from Springer Nature)

2.4.2 *Interfacial Thermal Transport*

Interfacial thermal resistance represents a dominant bottleneck in heat dissipation for nanoscale devices, particularly in heterostructures incorporating 2D materials such as graphene and MoS_2 on different substrates. Raman thermometry offers a unique capability to measure temperature differences across interfaces and quantify interfacial thermal resistance. For example, in the Raman-penetration method, an electrical current is passed through a supported 2D material (e.g., graphene), inducing steady-state Joule heating. A confocal Raman laser is then used to simultaneously excite and probe the sample, enabling distinct Raman peaks from the film and substrate to be resolved (Fig. 2.9). The temperature difference across the interface

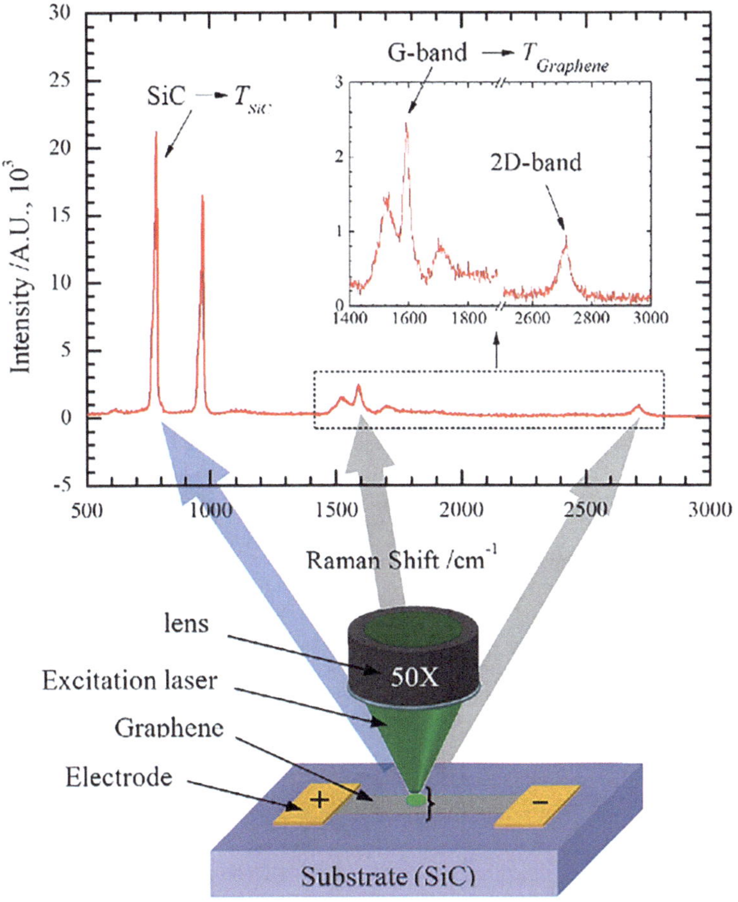

Fig. 2.9 The schematic of the electrical heating Raman-probing method to measure interfacial thermal contact resistance between graphene and 4H-SiC (reprinted from ref. [11], Copyright (2011), with permission from Wiley)

can be determined from these peaks, and the interfacial thermal resistance R_{int} can be calculated.

This method has revealed interfacial resistances in graphene/SiC systems. These Raman-based methods underscore how interfacial structure, contact quality, and material properties dictate nanoscale heat flow, providing valuable insights into the design of efficient thermal interfaces for advanced electronics. Detailed discussions on interfacial thermal transport and its implications will be presented in Chap. 5.

2.4.3 Sub-10 nm Thermal Probing

Near-field scanning optical microscopy (NSOM) enables the optical measurements below the 10 nm scale by confining laser energy to a nanoscale region using a sharp metallic tip [23–26]. As shown in Fig. 2.10, when the tip is brought into proximity with the sample, plasmonic resonance at the tip apex creates strong electromagnetic field enhancement [27, 28], which is known as tip-enhanced Raman scattering (TERS). This localized field enhancement amplifies both the incident laser intensity and the Raman-scattered signal, greatly improving spatial resolution and sensitivity of the signal.

With tip-enhancement effect, Raman can be used to detect temperature-induced spectral shifts within this localized volume, enabling nanoscale thermal probing [29]. By combining NSOM and Raman thermometry, simultaneous thermal probing and structural properties become possible, making these techniques indispensable tools for investigating heat dissipation in low-dimensional materials, plasmonic

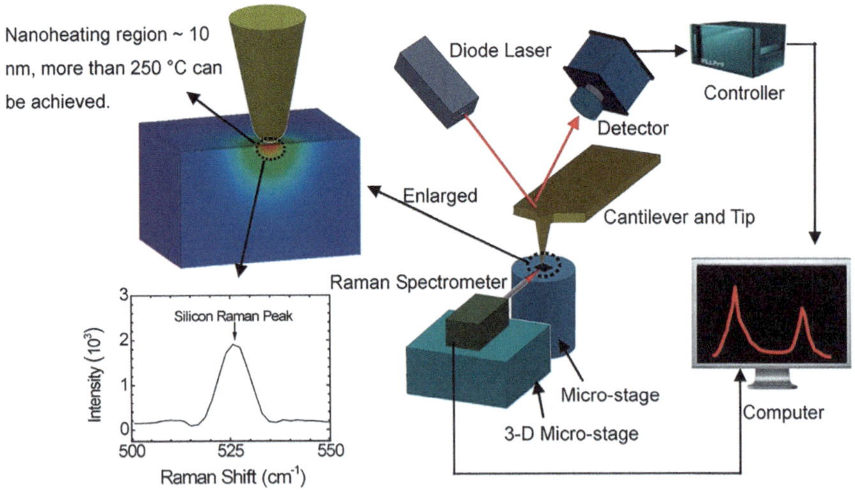

Fig. 2.10 Schematic of temperature measurement in near-field laser heating experiment (reprinted from ref. [29], with permission from Springer Nature)

devices, and other nanoscale systems. Further investigations and potential applications of sub-10 nm thermal probing will be introduced in Chap. 7.

2.4.4 Phonon Non-equilibrium

Raman spectroscopy provides unique access to phonon non-equilibrium phenomena in nanomaterials by probing the temperature of optical phonon (OP) modes through their characteristic spectral shifts. In suspended 2D materials such as MoS_2 and $MoSe_2$, Raman measurements under localized laser heating have revealed substantial temperature differences between optical and acoustic phonon (AP) populations. In graphene, for instance, the G-band and 2D-band Raman peaks correspond to distinct phonon modes, allowing independent tracking of OP and AP temperatures [30]. Studies have shown that overlooking this phonon non-equilibrium can lead to significant underestimation of thermal conductivity. Similar effects are observed in transition metal dichalcogenides like MoS_2, where Raman peak shifts under continuous-wave or nanosecond laser heating enable quantification of the energy coupling factor between OP and AP.

Phonon non-equilibrium arises when energy distribution among phonon modes deviates from thermal equilibrium, often under nanoscale excitations like focused laser irradiation or localized electrical bias. As shown in Fig. 2.11, such excitation can inject energy into specific modes, causing optical phonons to accumulate excess energy and exhibit a higher effective temperature than acoustic phonons. This imbalance hinders efficient energy dissipation, as acoustic phonons, which dominate heat conduction, cannot effectively transfer heat away from hotspots due to weak coupling with optical phonons. Thus, understanding and characterizing phonon non-equilibrium is crucial for advancing nanoelectronics in the post-Moore era. A thorough exploration of these phenomena will be detailed in Chap. 8.

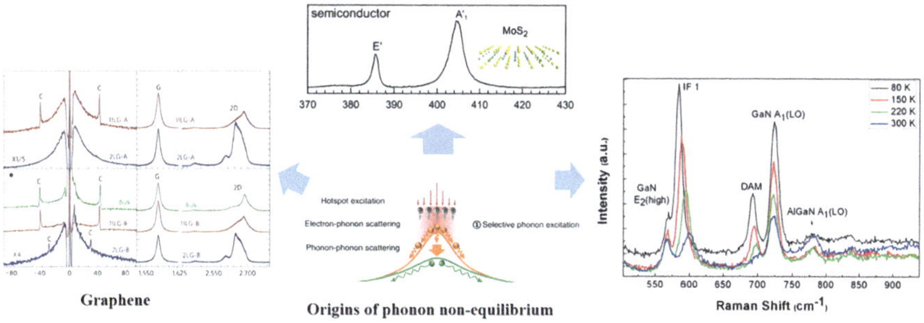

Fig. 2.11 Raman characterization of optical phonon modes (reprinted from ref. [30–33], with permission from Springer and American Physical Society)

2.5 Summary

This chapter has introduced the fundamental principles of Raman thermometry, highlighting the role of Raman spectroscopy in probing material properties, and focused on the temperature-dependent characteristics of Raman shift, line width, and intensity. These thermally sensitive features enable Raman thermometry to perform high-resolution, non-invasive temperature measurements at the micro/nanoscale, making it a valuable tool in the study of nanomaterials and nanoelectronic devices. Representative applications demonstrate the feasibility of this technique: measuring thermal conductivity of graphene, quantifying interfacial thermal resistance, sub-10 nm temperature measurement, and probing phonon non-equilibrium phenomena.

References

1. Raman, C.V., Krishnan, K.S.: A new type of secondary radiation. Nature. **121**(3048), 501–502 (1928)
2. Long, D.A.: Classical theory of Rayleigh and Raman scattering. In: The Raman Effect: A Unified Treatment of the Theory of Raman Scattering by Molecules, pp. 31–48 (2002)
3. Long, D.A.: Quantum mechanical theory of Rayleigh and Raman scattering. In: The Raman Effect: A Unified Treatment of the Theory of Raman Scattering by Molecules, pp. 49–84 (2002)
4. Smith, E., Dent, G.: Modern Raman Spectroscopy: a Practical Approach. John Wiley & Sons (2019)
5. Ferrari, A.C., Basko, D.M.: Raman spectroscopy as a versatile tool for studying the properties of graphene. Nat. Nanotechnol. **8**(4), 235–246 (2013)
6. Beams, R., Gustavo Cançado, L., Novotny, L.: Raman characterization of defects and dopants in graphene. J. Phys. Condens. Matter. **27**(8), 083002 (2015)
7. Kudelski, A.: Analytical applications of Raman spectroscopy. Talanta. **76**(1), 1–8 (2008)
8. Beechem, T., Graham, S.: Temperature Measurement of Microdevices Using Thermoreflectance and Raman Thermometry. In: BioNanoFluidic MEMS, pp. 153–174. Springer US, Boston, MA (2008)
9. Long D A. The Raman Effect: a Unified Treatment of Raman Scattering by Molecules. (2002)
10. Childs, P.R.N., Greenwood, J.R., Long, C.A.: Review of temperature measurement. Rev. Sci. Instrum. **71**(8), 2959–2978 (2000)
11. Yue, Y., Zhang, J., Wang, X.: Micro/nanoscale spatial resolution temperature probing for the interfacial thermal characterization of epitaxial graphene on 4H-SiC. Small. **7**(23), 3324–3333 (2011)
12. Tang, X., Xu, S., Wang, X.: Thermal probing in single microparticle and microfiber induced near-field laser focusing. Opt. Express. **21**(12), 14303–14315 (2013)
13. Tang, X., Xu, S., Wang, X.: Corrugated epitaxial graphene/SiC interfaces: photon excitation and probing. Nanoscale. **6**(15), 8822–8830 (2014)
14. Asher, S.A., Bormett, R.: Raman instrumentation. In: Raman Scattering in Materials Science, pp. 35–54. Springer Berlin Heidelberg, Berlin, Heidelberg (2000)
15. Wallis, R.F., Balkanski, M.: Many-Body Aspects of Solid State Spectroscopy. North-Holland Physics, Amsterdam (1986)
16. Rani, C., Tanwar, M., Ghosh, T., Kandpal, S., Saxena, S.K., Kumar, R.: Non-linear temperature dependent Raman parametric changes: an identification of Fano intervened systems. Phys. Rep. **1037**, 1–41 (2023)

17. Hart, T., Aggarwal, R., Lax, B.: Temperature dependence of Raman scattering in silicon. Phys. Rev. B. **1**(2), 638–642 (1970)
18. Iguo, J., Wang, X., Zhang, L., Wang, T.: Transient thermal characterization of micro/submicroscale polyacrylonitrile wires. Appl. Phys. A. **89**(1), 153–156 (2007)
19. Wang, T., Wang, X., Guo, J., Luo, Z., Cen, K.: Characterization of thermal diffusivity of micro/nanoscale wires by transient photo-electro-thermal technique. Appl. Phys. A. **87**(4), 599–605 (2007)
20. Sokół, A.K., Sarzała, R.P.: Thermal management of GaInNAs/GaAs VECSELs. Opto-Electron. Rev. **21**(2), 191–198 (2013)
21. Begtrup, G.E., Ray, K.G., Kessler, B.M., Yuzvinsky, T.D., Garcia, H., Zettl, A.: Probing nanoscale solids at thermal extremes. Phys. Rev. Lett. **99**(15), 155901 (2007)
22. Giberti, A., Guidi, V., Vincenzi, D.: A study of heat distribution and dissipation in a micromachined chemoresistive gas sensor. Sensors Actuators B Chem. **153**(2), 409–414 (2011)
23. Antunes, E.F., Lobo, A.O., Corat, E.J., Trava-Airoldi, V.J., Martin, A.A., Veríssimo, C.: Comparative study of first- and second-order Raman spectra of MWCNT at visible and infrared laser excitation. Carbon. **44**(11), 2202–2211 (2006)
24. Costa, S., Borowiak-Palen, E., Kruszyńska, M., Bachmatiuk, A., Kaleńczuk, R.J.: Characterization of carbon nanotubes by Raman spectroscopy. Materscipol. **26**(2), 433–441 (2008)
25. Freitag, M., Steiner, M., Martin, Y., Perebeinos, V., Chen, Z., Tsang, J.C., Avouris, P.: Energy dissipation in graphene field-effect transistors. Nano Lett. **9**(5), 1883–1888 (2009)
26. Huang, F., Yue, K.T., Tan, P., Zhang, S.-L., Shi, Z., Zhou, X., Gu, Z.: Temperature dependence of the Raman spectra of carbon nanotubes. J. Appl. Phys. **84**(7), 4022–4024 (1998)
27. Zhang, Y., Xie, L., Zhang, J., Wu, Z., Liu, Z.: Temperature coefficients of Raman frequency of individual single-walled carbon nanotubes. J. Phys. Chem. C. **111**(38), 14031–14034 (2007)
28. Zhou, Z., Dou, X., Ci, L., Song, L., Liu, D., Gao, Y., Wang, J., Liu, L., Zhou, W., Xie, S., Wan, D.: Temperature dependence of the Raman spectra of individual carbon nanotubes. J. Phys. Chem. B. **110**(3), 1206–1209 (2006)
29. Yue, Y., Eres, G., Wang, X., Guo, L.: Characterization of thermal transport in micro/nanoscale wires by steady-state electro-Raman-thermal technique. Appl. Phys. A. **97**(1), 19–23 (2009)
30. Tan, P.H., Han, W.P., Zhao, W.J., Wu, Z.H., Chang, K., Wang, H., Wang, Y.F., Bonini, N., Marzari, N., Pugno, N., Savini, G., Lombardo, A., Ferrari, A.C.: The shear mode of multilayer graphene. Nat. Mater. **11**(4), 294–300 (2012)
31. Cong, X., Liu, X.-L., Lin, M.-L., Tan, P.-H.: Application of Raman spectroscopy to probe fundamental properties of two-dimensional materials. NPJ 2D Mater. Appl. **4**(1), 13 (2020)
32. Liu, Y., Chen, D., Wei, G., Lin, Z., He, A., Li, M., Wang, P., Zhang, R., Zheng, Y.: Temperature-dependent ultraviolet Raman scattering and anomalous Raman phenomenon of AlGaN/GaN heterostructure. Opt. Express. **27**(4), 4781–4788 (2019)
33. Xu, J., Huang, X., Sheng, Y., Sun, Q., Zhang, H., Bao, H., Yue, Y. Direct Observation of Substantial Phonon Nonequilibrium Near Nanoscale Hotspots in Gallium Nitride. Adv. Sci. **12**(11), 2411040 (2025)

Chapter 3
Raman Characterization of Low-Dimensional Materials

Abstract The increasing role of low-dimensional materials in nanoscale devices makes it essential to understand their thermal conduction properties. However, their atomic-scale thickness and high interfacial thermal resistance pose significant challenges for conventional characterization. This chapter focuses on Raman-based steady-state methods for probing the thermal transport behavior of such materials. Section 3.1 outlines the major challenges in low-dimensional characterization. Section 3.2 introduces the steady-state electro-Raman thermal technique, which combines electrical heating and Raman thermometry to evaluate the thermal conductivity and interfacial resistance of nanoscale systems. Section 3.3 presents the steady-state laser heating Raman characterization approach, employing localized optical excitation to induce controlled temperature rises for non-contact thermal measurements.

Keywords Low-dimensional materials · Steady-state electro-Raman method · Laser heating characterization

3.1 Challenges in Low-Dimensional Characterization

Traditional thermal management strategies rely on heat sinks, cooling systems, and thermal interface materials with high thermal conductivity [1, 2]. The emergence of low-dimensional materials offers solutions for high thermal conductivity materials with less space needed. For example, carbon-based materials in two-dimensional (2D) form (graphene) or one-dimensional (1D) form (carbon nanotubes, CNTs) show strong phonon transport properties, which significantly increase their thermal conductivity, and often produce exceptional property compared with the bulk materials (Fig. 3.1) [3]. For instance, graphene is reported to have an intrinsic thermal conductivity of more than 1000 W/mK, and still maintains excellent thermal conductor when supported on a substrate, enabling efficient heat dissipation in transistor applications. Single CNTs reach thermal conductivity of more than 1000 W/mK,

Y. Yue, *Raman Thermometry*, https://doi.org/10.1007/978-981-95-5985-5_3

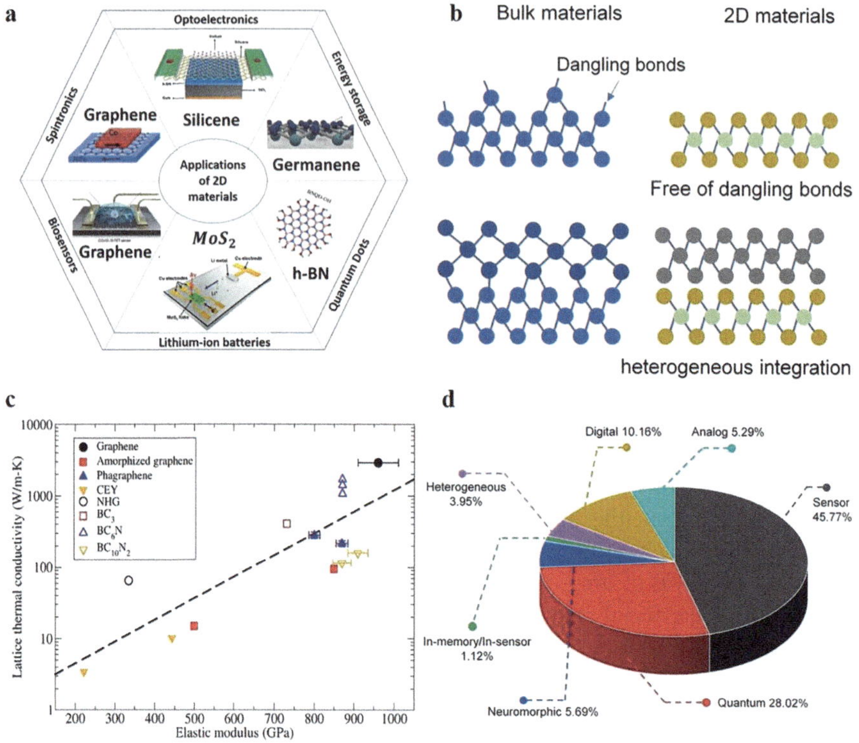

Fig. 3.1 (**a**) Engineering applications of low-dimensional materials. (**b**) Partial advantages of 2D materials over bulk materials [3]. (**c**) Excellent heat transfer properties of low-dimensional materials. (**d**) Number of 2D material publications from 2013 to 2023 [3]. (Reprinted with permission from Elsevier ref. [6, 7])

ranking among the highest thermal conductivity of all known materials. Beyond graphene and CNTs, many other low-dimensional materials, such as hexagonal boron nitride (*h*-BN), silicene, molybdenum disulfide (MoS_2), and black phosphorus, have emerged in recent years as potential candidates for next-generation materials [4, 5].

Despite their extraordinary thermal properties, the integration of low-dimensional materials presents significant challenges in terms of characterization. Accurate measurement of their thermal properties at the nanoscale is complicated by thermal resistance, structural defects, and fabrication constraints [8–10]. Due to its non-contact nature, Raman spectroscopy is often used for temperature measurement and heat transfer analysis. However, there are still problems with its application in the measurement of low-dimensional materials. The extremely low optical absorption and weak Raman signals of two dimensional (2D) materials make accurate temperature extraction difficult. The laser itself can induce significant local heating, resulting in uncertain temperature calibration and strongly localize temperature gradients that deviate from the uniform heating assumptions required for ideal conditions. The thermal resistance between material and substrate further complicates the thermal measurement process, as some of the heat may dissipate through the interface rather than in the plane direction. Furthermore, temperature fields in suspended structures are significantly affected by air convection and nanoscale radiation, which are often neglected in traditional Raman analysis. The combined effects of low absorption, interface resistance, laser-induced heating and convection result in substantial uncertainty and complexity when interpreting Raman thermal measurements for low-dimensional materials.

It is difficult for both steady-state and transient methods to measure temperature at atomic resolution in 2D materials. One approach to address this has been to increase the measurable thickness by depositing an additional layer atop the 2D material, allowing the use of conventional methods. However, such modification alters phonon transport by introducing constraint effects that affect the intrinsic behavior. The natural corrugation, wrinkling, and grain boundaries present in many 2D materials introduce additional complexity into experimental analysis [11–13]. For 1D materials such as CNTs or nanowires, the extremely small cross-sectional area further weakens the electrical or optical signal. Understanding the mechanisms of heat transfer across such atomic-scale materials is therefore essential for optimizing device-level thermal performance. In response to these challenges, the following sections will briefly introduce the Raman-based techniques developed in the PI's lab designed for characterizing thermal property of low-dimensional materials.

3.2 Steady-State Electro-Raman Thermal Technique

Raman has emerged as an ideal non-contact approach for temperature probing because Raman spectra (including wavenumber, linewidth, and intensity) are sensitive to temperature variations [14]. This technique has been successfully applied to a variety of materials such as silicon and CNTs to study the temperature variation [15]. In this context, the steady-state electro-Raman thermal (SERT) technique is developed to measure thermal properties of micro- and nanowires (representing 1D structures).

The SERT method involves two steps: calibration and thermal characterization. In the calibration step, the sample (e.g. the CNT fiber or bundle) is placed on a heating plate whose temperature is kept uniform and constant. Since the plate and the sample remain in close contact, they can be considered to share the same temperature. Raman measurements of the sample are taken at several temperatures, and the relation between Raman peak and temperature is obtained.

In thermal characterization, the sample (e.g the CNT fiber) is fabricated to be suspended across two electrodes, as shown in Fig. 3.2a. A direct current (DC) power supply is applied to the sample to generate electric heating. Based on the one-dimensional heat transfer equation, the maximum temperature occurs at the midpoint of the sample. This value can be greatly affected by both the thermal

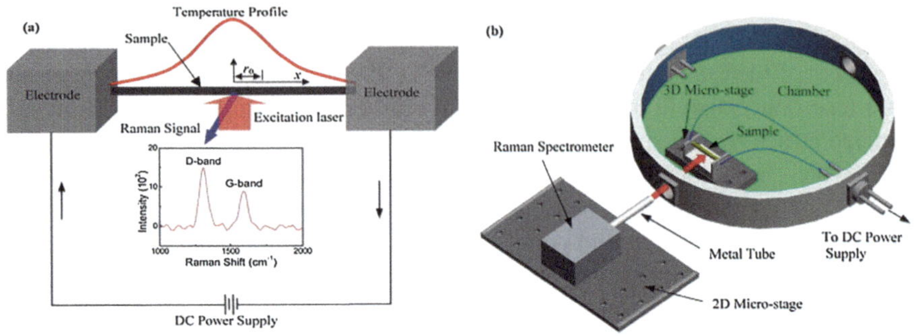

Fig. 3.2 (**a**) Schematic of the experimental principle for the SERT technique. (**b**) Schematic of experimental setup. (Reprinted with permission from Elsevier ref. [14])

conductivity and the input power. The same laser used in the calibration stage is employed again to excite the sample for Raman spectroscopy, enabling the monitoring of its temperature changes. After establishing the relationship between temperature and applied DC power, the thermal conductivity can be derived through linear fitting [14].

Figure 3.2a shows a conductive micro/nanowire with a total length of 2 L suspended between two electrodes. The wire axis is taken as the x-direction, and the midpoint of the sample is set as $x = 0$. When a DC current is passed through the sample, the sample will be heated up due to the joule effect. Because the electrodes are excellent heat sinks due to their large size compared with the sample, the ends of the wire stay at room temperature (T_0). During the experiment, the wire is placed in a vacuum and protected with aluminum foil, which means that only heat conduction along the wire is important while the heat convection and thermal radiation can be negligible. The temperature distribution along the wire can be described as $T(\mathrm{x}) = I^2 \cdot R \cdot L/(4\,k \cdot A_c) \cdot (1 - x^2/L^2) + T_0$, where R, k, A_c are the electrical resistance, thermal conductivity, and the cross-sectional area of the wire, respectively. At midpoint, the temperature is $T(0) = I^2 \cdot R \cdot L/(4\,k \cdot A_c) + T_0$. Theoretically, by only measuring one midpoint temperature and the input power, the thermal conductivity of the sample can be determined. In the measurement, to suppress the uncertainty, different heating powers are used and the corresponding temperature are measured. The relationship between the heating power and midpoint temperature can be fitted linearly to determine the thermal conductivity of the wire.

In the characterization, it is a concern that the heating effect from the probing laser would introduce some uncertainty. In ref. [14], there is detailed mathematical derivation for accounting such effect: for example, if a DC current I pass through the wire, creating steady Joule heating by $Q_e = I_2R/(A_c \cdot L)$, with R being the resistance of the sample. A probing laser with a spot size of diameter $2r_0$ is applied to the wire, leading to an absorbed energy density of $Q_p = Q_p/(A_c \cdot r_0)$, where Q_p represents the total absorbed laser energy and A_c is the cross-sectional area of the wire. If considering that the thermal conductivity of of the wire is temperature dependent: Umklapp phonon–phonon scattering introduces a $1/T$ dependence, and three-phonon scattering contributes a second-order dependence. The thermal conductivity is written follows the relation $1/(\alpha T + \beta T^2)$ [16–19]. If the temperature increase is small (e.g., less than 100 K above room temperature), the first-order term dominates [20]. The thermal conductivity can be approximated as $\alpha/(T + \beta)$, where α and β are constants. The temperature profile of the wire in the region $x = 0$ to r_0 can then be expressed as:

$$T(x) = (T_0 + \beta)\cdot \exp\left(-\frac{\dot{Q}_e + \dot{Q}_p}{2\alpha}x^2\right) - \beta\,(0 < x < r_0) \tag{3.1}$$

The connection between the midpoint temperature (T_0) and the endpoint temperature (T_L) is given by:

$$T_0 = (T_L + \beta)\cdot \exp\left(-\frac{\dot{Q}_p}{2\alpha}r_0^2 + \frac{\dot{Q}_e}{2\alpha}L^2 + \frac{\dot{Q}_p}{\alpha}r_0\cdot L\right) - \beta \tag{3.2}$$

The average temperature of the region subjected to laser illumination on the sample is formulated as:

$$\bar{T}_{r,c} = (T_L + \beta)\cdot \exp\left(-\frac{\dot{Q}_p}{2\alpha}r_0^2 + \frac{\dot{Q}_p}{\alpha}r_0\cdot L\right) \cdot \frac{1}{2r_0}\sqrt{\frac{2\pi\alpha}{\dot{Q}_p}}\cdot erf\left(r_0\sqrt{\frac{\dot{Q}_p}{2\alpha}}\right) - \beta \tag{3.3}$$

The endpoint temperature (T_L) is assumed to be equal to the hot plate temperature, which is recorded during the calibration stage. During the electrical heating experiments, the endpoint is regarded as constant at room temperature ($T_L = T_\infty$) because the electrodes act as heat sinks compared with the sample. In these measurements, although the electrical and laser heating powers are of similar order, the power density from electrical heating is much smaller than that of the laser ($\dot{Q}_e \ll \dot{Q}_p$) since the wire length is much greater than the laser spot size. Because the measurement temperatures (T_m) are obtained from the calibration results, the actual measured temperature should be expressed as:

$$T_m = (T_\infty + \beta)\cdot \exp\left(\frac{\dot{Q}_e \cdot L^2}{2\alpha}\right) - \beta \tag{3.4}$$

In addition, the correlation between T_m and T_∞ can be regarded as linear when the heating power density starts from zero, that is $T_m = Q_e L^2 (T_\infty + \beta)/2\alpha + T_\infty$. By substituting $k = \alpha/(T + \beta)$ and $\dot{Q}_e = I^2R/(Ac \bullet L)$ into the equation, the relationship between the measured temperature and the electrical heating power becomes

$T_m = PL/(4kA_c) + T_\infty$, where $P = I^2R$ is the input electric power and k is the thermal conductivity at room temperature. In principle, the thermal conductivity of the sample can be determined by linear fitting of the $T_m - P$ relation.

To demonstrate the capability of the SERT technique, the multiwalled carbon nanotube (MWCNT) buckypaper sample (cut as a strip) was examined. The measurement was carried out in a vacuum chamber with a pressure below 1×10^{-3} Torr, and the schematic of the system is shown in Fig. 3.2b. The MWCNT buckypaper was suspended between two copper electrodes on a sample holder, with the ends fixed by silver paste. The holder was placed on a 3D microstage, and the electrodes were connected to the electrical feedthrough of the chamber. Representative spectra obtained from the calibration experiments at different temperatures are presented in Fig. 3.3. The D-band peak was selected for analysis because it is sharper and has higher intensity compared to the G-band. Moreover, the spectra collected in this work shows distinct Raman peaks [21]. The thermal conductivity is measured as 2.92 W/(mK), which is attributed to the tangled CNT structure and the low thermal conductance of the CNT–CNT contact [14].

In addition to the above analysis for the uncertainty of SERT technique, the electrical contact resistance between the sample and electrodes can cause undesired effects, especially for samples with low intrinsic resistance. During calibration, the entire sample and contact points reach the same temperature. Actual thermal conductivity measurements heat only the sample, leaving the contact points near room temperature. This difference can significantly affect measurements based on electrical resistance and may even cause them to fail. In above experiment, laser is focused on the midpoint of the sample, where the effect of contact resistance is minimal. For

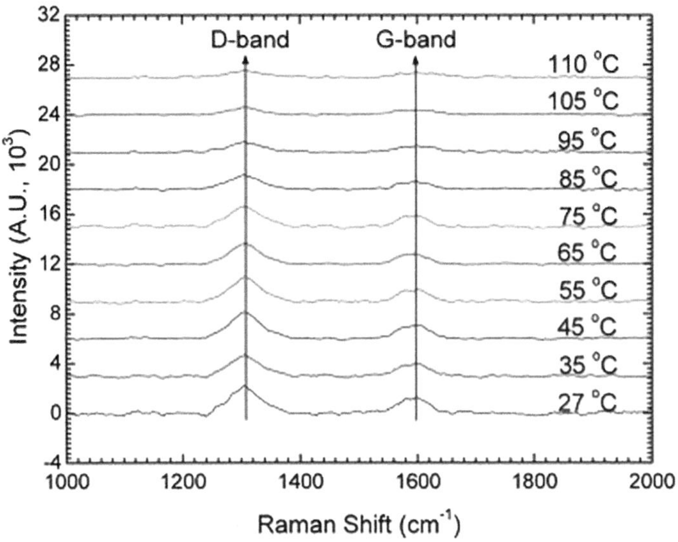

Fig. 3.3 Spectra at different temperatures in the calibration. (Reprinted with permission from Elsevier ref. [14])

samples much shorter than the laser spot size, the SERT technique can be used to measure the average sample temperature rather than the midpoint. This adaptation makes suitable for evaluating the thermal conductivity of much shorter micro/nanowires.

3.3 Steady-State Laser Heating Raman Characterization

In SERT technique, the laser-induced heating is minimized or corrected to avoid bias in the thermal conductivity measurement. Sometimes, if the focused laser spot and intense photon flux cause significant temperature rise, or if we want to apply the laser heating effect, the absorbed laser energy dominates heat dissipation in the illuminated region [22]. On the contrary, it allows direct study of thermal transport from a point heat source. In the following, the SERT technique has been modified so that Joule heating is eliminated and laser-induced heating is the only source of heat. This modification avoids the need for nanofabricated electrical circuits, greatly simplifying sample preparation. It also works for samples down to sub-micron lengths while keeping measurements fast and straightforward.

The experimental principle is similar to the SERT technique. For micro/nanowires the sample is bridge-suspended on two bulk heat sinks and has its ends firmly attached. The absorbed laser energy heats the irradiation spot, and the heat then transfers along the wire toward the two ends. Since the ends are connected to bulk materials with large heat capacity, their temperatures can be considered close to room temperature. The corresponding Raman signal is collected during the laser heating process. Raman thermometry provides the temperature at the sample midpoint. This measurement establishes the relationship between heating power and temperature rise. Figure 3.4 illustrates the experimental setup for measuring the

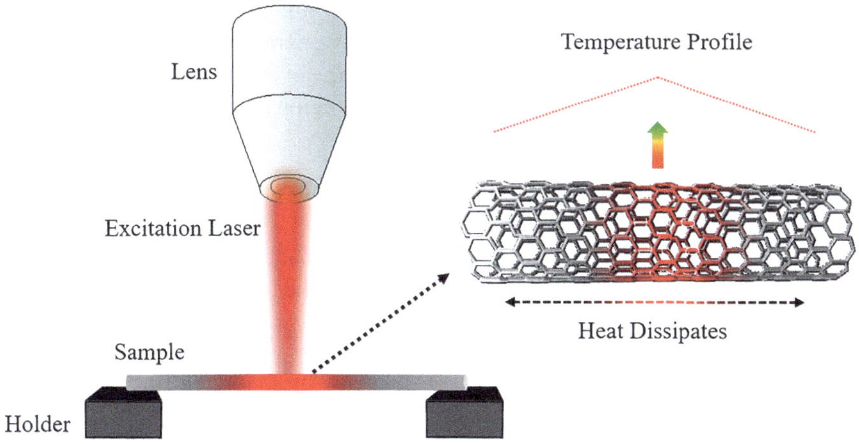

Fig. 3.4 Schematic of the experimental principle for thermal property measurements of micro/nanowires

thermal properties of micro/nanowires. Assuming that the thermal properties remain constant within a narrow temperature range, the governing equation of this 1D steady-state model in the laser-irradiated region is $k\Delta T + \dot{q} = 0$. where T is the temperature, k is the thermal conductivity of the material, and $\dot{q}$ is the effective volumetric heat generation rate (W/m^3) in the sample. The heat generation term $\dot{q}$ can be derived from the absorbed laser energy Q as $\dot{q} = Q / 2\pi r_s^2 r_0$, where r_s is the sample radius and r_0 is the half-width of the laser beam. Outside the irradiated region, the governing equation simplifies to one without internal heat generation, expressed as $-kA\Delta T = Q/2$, since the heat is conducted symmetrically toward both ends.

Within the laser irradiated region ($r < r_0$, r_0 denoting the half-length of the laser spot), the heat conduction equation is solved together with the no irradiated region. The temperature distribution can be obtained by applying boundary conditions where the sample ends are at room temperature and there is no heat flux at r at $r = 0$ as:

$$T(r) = -\frac{Q}{4\pi k r_0 r_s^2} r^2 + \frac{Q}{2\pi k r_s^2}\left(L - \frac{r_0}{2}\right) + T_0 \tag{3.5}$$

The average temperature within the laser irradiation region can be derived as:

$$\bar{T} = \frac{1}{r_0}\int_0^{r_0} T(r)\,dr = \frac{QL}{2\pi k r_s^2} - \frac{Q r_0}{3\pi k r_s^2} + T_0 \tag{3.6}$$

It is shown that the average temperature within the laser-heated region exhibits a linear dependence on the absorbed laser energy, with the proportionality coefficient given by $\left(L / 2\pi r_s^2 - r_0 / 3\pi r_s^2\right) / k$. Since the temperature can be directly determined from the Raman signal, the thermal conductivity of the micro/nanowires can be extracted.

Similar to the case of micro/nanowires, Fig. 3.5 illustrates the measurement principle used for the thermal characterization of thin films (2D structure). Assuming that the thermal properties remain constant within a narrow temperature range and considering only steady-state heat conduction, the governing equation for the 2D heat dissipation model within the laser-irradiated region can be expressed as:

$$k\frac{1}{r}\frac{d}{dr}\left(r\frac{dT}{dr}\right) + \dot{q} = 0 \tag{3.8}$$

where T denotes the temperature, r represents the radial distance from the center of the sample, k is the thermal conductivity, and $\dot{q}$ is the equivalent volumetric heat generation rate defined as $\dot{q} = Q / \pi r_0^2 d$, where r_0 is the equivalent radius of the laser irradiated area and d is the sample thickness. Outside the laser heating region, the heat transfer is governed by $-kAdT/dr = Q$, where r is the radial position in the sample and A is the effective cross-sectional area through which heat dissipates

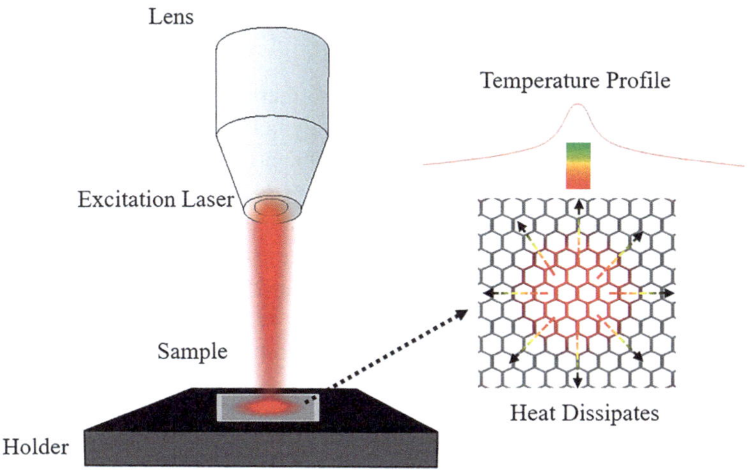

Fig. 3.5 Schematic of the experimental principle for thermal property measurements of thin films

from the center to the edge. For the region $r < r_0$, temperature distribution can be derived as:

$$T(r) = \frac{Q}{4k\pi r_0^2 d}\left(r_0^2 - r^2\right) + \frac{Q}{2k\pi d}\left(\ln r_L - \ln r_0\right) + T_0 \tag{3.9}$$

The average temperature in laser irradiation area is:

$$\overline{T} = \frac{1}{\pi r_0^2}\int_0^{2\pi} d\varphi \int_0^{r_0} rT(r)\,dr = \frac{Q}{k}\left[\frac{l}{8\pi d} + \frac{l}{2\pi d}\left(\ln r_L - \ln r_0\right)\right] + T_0 \tag{3.10}$$

The average temperature in the laser-irradiated region exhibits a linear dependence on the absorbed laser energy Q, with the proportionality coefficient given by $[(\ln r_L - \ln r_0)/2 + 1/8]/\pi dk$. The temperature can be directly extracted from the Raman signal, allowing the thermal conductivity of the sample to be determined.

This laser-heating Raman technique is particularly suitable for thermal characterization of extremely small samples, such as those with sub-micron dimensions. The laser beam can fully cover the sample, and the measured temperature corresponds to the mean value. Assuming an absorbed laser intensity $\overline{q}$, the mean temperature for micro/nanowires is expressed as $\overline{T} = \dot{q}L^2 / 3k + T_0$, where L is half the wire length. The slope ($L^2/3k$) of the linear relation between heating power and mean temperature provides the thermal conductivity, which slightly differs from bulk materials. For thin films with dimensions below 1 μm, the mean temperature is $\overline{T} = \overline{q}r_L^2 / 8k + T_0$, where r_L represents the effective sample radius (e.g., for a circular sample). The slope ($r^2_L/8\,k + T_0$) can then be used to determine the thermal conductivity of the film.

3.4 Summary

This chapter introduced the challenges of thermal characterization in low-dimensional materials and presented two steady-state Raman-based thermal measurement methods: one employing electro-Raman-thermal heating and the other utilizing laser-induced heating. Both approaches extract steady-state temperature rises from the Raman scattering signals of the materials and determine thermal conductivity through the heat transfer models. Compared with contact-based techniques, these methods enable non-contact operation with high spatial resolution, avoiding interfacial resistance. Since the Raman scattering signal originates from the measured target, the techniques enable temperature probing at small scales and provide direct insights into the thermal properties of low-dimensional materials.

References

1. Ma, H., Gao, B., Wang, M., Yuan, Z., Shen, J., Zhao, J., Feng, Y.: Strategies for enhancing thermal conductivity of polymer-based thermal interface materials: a review. J. Mater. Sci. **56**(2), 1064–1086 (2021)
2. Xing, W., Xu, Y., Song, C., Deng, T.: Recent advances in thermal interface materials for thermal management of high-power electronics. Nanomaterials. **12**(19), 3365 (2022)
3. Liu, A., Zhang, X., Liu, Z., Li, Y., Peng, X., Li, X., Qin, Y., Hu, C., Qiu, Y., Jiang, H.: The roadmap of 2D materials and devices toward chips. Nano-Micro Lett. **16**(1), 119 (2024)
4. Upadhyay, S.N., Satrughna, J.A.K., Pakhira, S.: Recent advancements of two-dimensional transition metal dichalcogenides and their applications in electrocatalysis and energy storage. Emerg. Mater. **4**(4), 951–970 (2021)
5. Ranjan, P., Gaur, S., Yadav, H., Urgunde, A.B., Singh, V., Patel, A., Vishwakarma, K., Kalirawana, D., Gupta, R., Kumar, P.: 2D materials: increscent quantum flatland with immense potential for applications. Nano Convergence. **9**(1), 26 (2022)
6. Gautam, S., Singhal, P., Chakraverty, S., Goyal, N.: Synthesis and characterization strategies of two-dimensional (2D) materials for quantum technologies: a comprehensive review. Mater. Sci. Semicond. Process. **181**, 108639 (2024)
7. Pereira, L.F.C.: Investigating mechanical properties and thermal conductivity of 2D carbon-based materials by computational experiments. Comput. Mater. Sci. **196**, 110493 (2021)
8. Shi, L.: Thermal and thermoelectric transport in nanostructures and low-dimensional systems. Nanoscale Microscale Thermophys Eng. **16**(2), 79–116 (2012)
9. El Sachat, A., Alzina, F., Sotomayor Torres, C.M., Chavez-Angel, E.: Heat transport control and thermal characterization of low-dimensional materials: a review. Nanomaterials. **11**(1), 175 (2021)
10. Paras, Yadav, K., Kumar, P., Teja, D.R., Chakraborty, S., Chakraborty, M., Mohapatra, S.S., Sahoo, A., Chou, M.M., Liang, C.-T.: A review on low-dimensional nanomaterials: nanofabrication, characterization and applications. Nanomaterials. **13**(1), 160 (2022)
11. Yao, W., Wu, B., Liu, Y.: Growth and grain boundaries in 2D materials. ACS Nano. **14**(8), 9320–9346 (2020)
12. Chen, W., Gui, X., Yang, L., Zhu, H., Tang, Z.: Wrinkling of two-dimensional materials: methods, properties and applications. Nanoscale Horiz. **4**(2), 291–320 (2019)
13. Androulidakis, C., Zhang, K., Robertson, M., Tawfick, S.: Tailoring the mechanical properties of 2D materials and heterostructures. 2D Materials. **5**(3), 032005 (2018)

14. Yue, Y., Huang, X., Wang, X.: Thermal transport in multiwall carbon nanotube buckypapers. Phys. Lett. A. **374**(40), 4144–4151 (2010)
15. Zhou, Z., Dou, X., Ci, L., Song, L., Liu, D., Gao, Y., Wang, J., Liu, L., Zhou, W., Xie, S., Wan, D.: Temperature dependence of the Raman spectra of individual carbon nanotubes. J. Phys. Chem. B. **110**(3), 1206–1209 (2006)
16. Pop, E., Mann, D., Cao, J., Wang, Q., Goodson, K., Dai, H.: Negative differential conductance and hot phonons in suspended nanotube molecular wires. Phys. Rev. Lett. **95**(15), 155505 (2005)
17. Mingo, N., Broido, D.A.: Length dependence of carbon nanotube thermal conductivity and the "problem of long waves". Nano Lett. **5**(7), 1221–1225 (2005)
18. Begtrup, G.E., Ray, K.G., Kessler, B.M., Yuzvinsky, T.D., Garcia, H., Zettl, A.: Probing nanoscale solids at thermal extremes. Phys. Rev. Lett. **99**(15), 155901 (2007)
19. Singh, T.J., Verma, G.S.: Resonance effects in three-phonon processes. Phys. Rev. B. **25**(6), 4106–4109 (1982)
20. Pop, E., Mann, D., Wang, Q., Goodson, K., Dai, H.: Thermal conductance of an individual single-wall carbon nanotube above room temperature. Nano Lett. **6**(1), 96–100 (2006)
21. Yue, Y., Eres, G., Wang, X., Guo, L.: Characterization of thermal transport in micro/nanoscale wires by steady-state electro-Raman-thermal technique. Appl. Phys. Mater. Sci. Process. **97**(1), 19–23 (2009)
22. Li, M., Yue, Y.: Raman-based steady-state thermal characterization of multiwall carbon nanotube bundle and Buckypaper. J. Nanosci. Nanotechnol. **15**(4), 3004–3010 (2015)

Chapter 4
Raman Measurement of Convective Heat Transfer

Abstract With device dimensions shrinking to the micro- and nanoscale, convective heat transfer at interfaces plays an increasingly important role in determining the overall thermal performance. Unlike bulk systems dominated by conduction, suspended or ultrathin structures interact strongly with surrounding air or gas molecules, introducing complex heat exchange mechanisms that are difficult to characterize accurately. This chapter presents Raman-based methods for quantifying convective heat transfer across micro- and nanoscale systems, ranging from suspended microwires to two-dimensional (2D) materials and nanoscale hotspots. Section 4.1 introduces the fundamentals of convective heat transfer. Section 4.2 details the steady-state Raman mapping method, which enables characterization of convective heat transfer coefficient and thermal conductivity of the wire simultaneously. Section 4.3 focuses on characterizing convective effects in 2D structures, while Sect. 4.4 explores heat convection at nanoscale hotspots, where localized heating effect drive unique air–solid interactions.

Keywords Convective heat transfer · Raman mapping · Low-dimensional structures · Nanoscale hotspots

4.1 Convective Heat Transfer in Low-Dimensional Materials

Although heat transfer across solid-solid interfaces has received the most attention, convective heat exchange between air molecules and solid surfaces is often viewed as less significant. However, for 2D materials, such as graphene and transition metal dichalcogenides (TMDs), the convective heat transfer can no longer be overlooked [1]. Especially for TMD materials, such as molybdenum disulfide (MoS_2), which have lower thermal conductivity than graphene, their interactions with air molecules play a more noticeable role in heat exchange (see Fig. 4.1). The heat transfer takes place over all exposed surfaces and has a substantial influence on the thermal management of 2D structures. This is especially true in suspended configurations, where these materials are used in devices including gas sensors, field emitters, and

Y. Yue, *Raman Thermometry*, https://doi.org/10.1007/978-981-95-5985-5_4

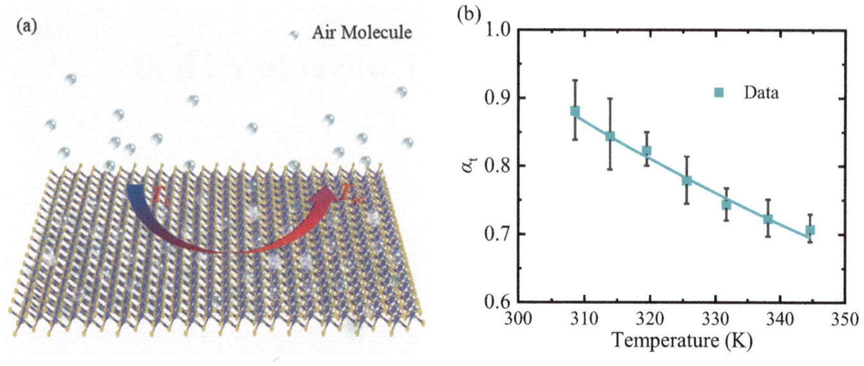

Fig. 4.1 (**a**) Schematic illustration of air molecule collisions on the MoS_2 surface. (**b**) Temperature dependence of the thermal accommodation coefficient. (Reprinted with permission from American Institute of Physics ref. [4])

detectors [2, 3]. Heat exchange between air molecules and 2D structures should be well understood, which is crucial for improving thermal management.

Previous studies (Chap. 3) used the SERT method, where the midpoint temperature of a sample is probed using Raman spectroscopy under steady-state electrical currents [5]. The thermal conductivity is derived from the linear correlation observed between Joule heating and the recorded midpoint temperature. Although this technique works well for measuring heat conduction in materials like carbon nanotube bundles and buckypapers, it usually demands a high-vacuum setting. The need for ultra-high vacuum, along with challenges involved in manipulating samples within a limited space, restricts the practicality of this approach. The influence of convection at micro- and nanoscale levels, which is particularly important for suspended 2D materials under ambient conditions, has been neglected. These constraints underscore the necessity for new measurement approaches that can accurately assess convective heat transfer in environments.

Heat transfer between air molecules and microscale solids is mainly driven by collisions between gas molecules and solid surfaces. These interactions are measured using the thermal accommodation coefficient (TAC), defined as the ratio of actual heat transfer to the maximum theoretically possible under the second law of thermodynamics. Although TAC values have been experimentally established for many gases and bulk materials, the behavior of 2D materials is more significantly affected by surrounding gases due to their exceptionally high surface to volume ratios. As structural dimension of devices shrinks to the nanoscale, the direct measurement of heat transfer at gas-solid interfaces becomes increasingly difficult.

The difficulty for measuring convective heat transfer between 2D materials and air is mainly due to their very thin structure and complex surface interactions [6]. However, there are a lot of applications (including thermoelectric devices, thermal interface materials, and solar cells) in air environments where both conduction and convection play a role [7]. As mentioned above for the SERT technique, non-contact optical methods such as Raman-based thermal measurements provide high spatial resolution and enable temperature measurement of low-dimensional (LD) materials. These capabilities make such techniques especially useful for studying heat exchange between air and LD materials. The following sections will introduce and examine several experimental methods designed to quantify convective heat transfer in LD materials under ambient conditions.

4.2 Steady-State Raman Mapping Method

To simultaneously measure both conduction and convection of 1D structure, a steady-state Raman mapping method was developed. This technique collects temperature data at multiple points while mapping the sample, producing temperature profiles instead of the single midpoint measurement used in the SERT method. These profiles are analyzed using a steady-state heat conduction model that includes a convective heat transfer term, allowing both the thermal conductivity (k) and the convection coefficient (h) to be determined at the same time for micro- and nanoscale wires [8]. This method is similar to the SERT technique and retains steady-state operation, non-contact sensing, and non-destructive testing.

During the measurement, the wire is suspended between two electrodes, and is heated by applying different constant currents. The ends of the sample are regarded to stay at room temperature due to the high thermal conductivity and large heat capacity of the electrodes. Heat is dissipated primarily through conduction into the electrodes and convection into the air. This process is described by the equation: $d^2T/dx^2 + (Q - Q_{air})/kAL = 0$, where T is the temperature, x is the distance from the midpoint, Q is the Joule heating power, k is the thermal conductivity, A is the cross-sectional area, and L is the length of the sample. The term Q_{air} represents convective heat loss and is given by: $Q_{air} = hLS(T - T_0)$, in which h is the convection coefficient, S is the perimeter of the wire, and T_0 is room temperature. By solving the equation under the corresponding boundary conditions, the steady-state temperature distribution can be obtained as:

$$T(x) = \frac{Q}{hLS}\left(1 - \frac{e^{\sqrt{hS/kA}x} + e^{-\sqrt{hS/kA}x}}{e^{\sqrt{hS/kA}L/2} + e^{-\sqrt{hS/kA}L/2}}\right) + T_0. \tag{4.1}$$

If the different temperatures along the wire are available when supplying with different heating powers, the thermophysical properties such as thermal conductivity and convective heat transfer coefficient can be obtained simultaneously by fitting the temperature-power relationship. To evaluate this method, a carbon nanotube (CNT) fiber with a diameter of 36 μm was characterized, and the temperature-dependent thermal properties, including thermal conductivity and convection coefficient, were measured in ambient air. Figure 4.2a shows a schematic of the experimental setup. Local temperature was measured by using distinctive Raman peaks of the CNT. During the measurement, the Raman peak can be fitted with a

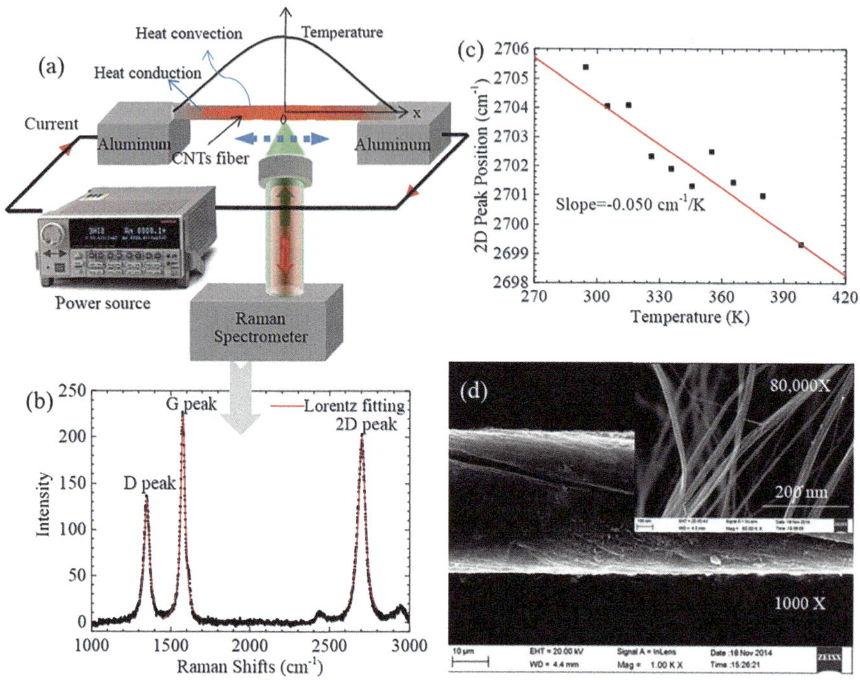

Fig. 4.2 (**a**) Schematic of the steady-state Joule-heating Raman mapping technique. (**b**) Raman spectrum of the CNT fiber at room temperature. (**c**) Linear relationship between the 2D peak positions and temperature. (**d**) SEM images showing the structure of the CNT fiber. (Reprinted with permission from American Institute of Physics ref. [8])

Lorentzian function (as displayed in Fig. 4.2b) to reduce the uncertainty. In this work, the temperature sensitivity of the 2D peak was 0.050 cm^{-1}/K, as shown in Fig. 4.2c. Scanning electron microscopy (SEM) images of the CNT fiber (presented in Fig. 4.2d) reveal its morphology and hierarchical structure. The high length to diameter ratio (3.0 mm to 36 μm) ensures that the one-dimensional thermal model is applicable.

We measured the convective heat transfer coefficient for different heating powers (temperatures). As shown in Fig. 4.3, the convection coefficient decreases steadily from 1143 W/m^2K to 1039 W/m^2K as the temperature rises from 335 to 468 K. The measured convection coefficients are significantly higher than those typically found in large-scale scenarios, where natural convection values generally range from 0 to 10 W/m^2K. This difference arises because heat convection exhibits behavior closer to conduction when the size of the fiber is decreased. Based on the empirical equation for microscale space $h = \left(2.68 + 0.11/\sqrt{D}\right)^2$ for 10 μm < D < 10 m [9], which was extensively adopted to analyze the microscale heat convection problem [10, 11], it is calculated as 274 W/m^2K. Given that the surface microstructure of the

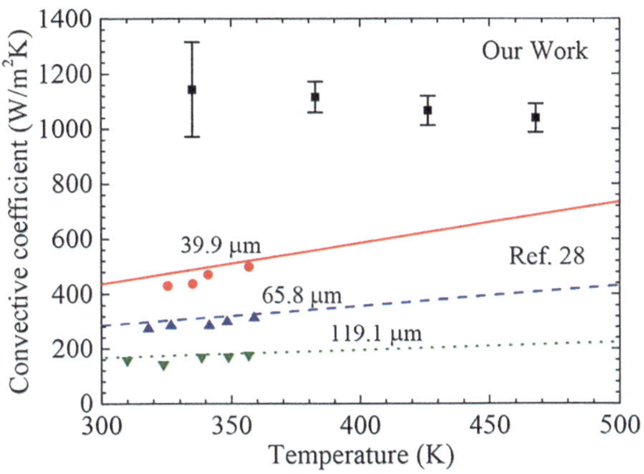

Fig. 4.3 Temperature dependence of the convection coefficient for the CNT fiber in air. (Reprinted with permission from American Institute of Physics ref. [8])

sample can significantly increase the total surface area, our measured value of 1100 W/m^2K is consistent with theoretical predictions.

Similar findings have been reported in other works: convection coefficients were measured at 400–500 W/m^2K for a copper microwire (39.9 μm) [12] and around 400 W/m^2K for a platinum microwire (32.6 μm) [13]. These exceptionally high values of h highlight the significant role of convective heat dissipation. Convective dissipation can be quantitatively estimated from the measured temperature profile and the fitted h value. If a heating power is 97 mW, the convective heat loss is calculated as: $\hat{Q}_{\text{air}} = \int \pi Dh\left(T - T_0\right)dx = 56.4$ mW, accounting for approximately 60% of the total energy input.

In addition, it is analyzed that the temperature dependence of convection coefficient arises from two main factors: (1) the direct influence of temperature, which alters physical properties of both the air and the fiber; and (2) indirect temperature effects, including thermal expansion and possible matrix melting during heating, which modify the internal microstructure of the material.

4.3 Characterization of Convective Heat Transfer in 2D Structures

The steady-state Raman mapping method can be extended to two-dimensional (2D) atomic-layer materials. Heat dissipation in these structures differs because of their ultra-thin nature and large exposed surface areas. The following work introduces a study using steady-state laser heating Raman probing method to evaluate convective heat transfer in suspended MoS_2 flake. To evaluation the effect of heat convection effect, we conducted two measurements: one is in vacuum and the other one in ambient air. By comparing the results from these measurements, the equivalent heat transfer coefficient (h_c) of the suspended MoS_2 flake can be obtained.

Here are the experimental details: the temperature increase of the MoS_2 flake mainly stems from the heat conduction effect under vacuum conditions, where the thermal conductivity (k) is important. Figure 4.4a displays the steady-state temperature distribution during continuous laser heating. Figure 4.4b shows the corresponding temperature profiles across the suspended region for different values of thermal conductivity. In the experiment, the thermal conductivity is determined by matching the experimental temperature rise with simulated results. It is noted that the observed temperature rise in the MoS_2 flake corresponds to an intensity-weighted average over the heated region. As shown in Fig. 4.4c, the calculated thermal conductivities for the MoS_2 flake decrease from 32.5 W/mK to 28.8 W/mK as the temperature increases from 310 K to 354 K. The thermal conductivity follows an approximately inverse relationship with temperature, expressed as: $k \propto T^{-1}$, which is caused by enhanced Umklapp phonon-phonon scattering at higher temperatures [7, 14].

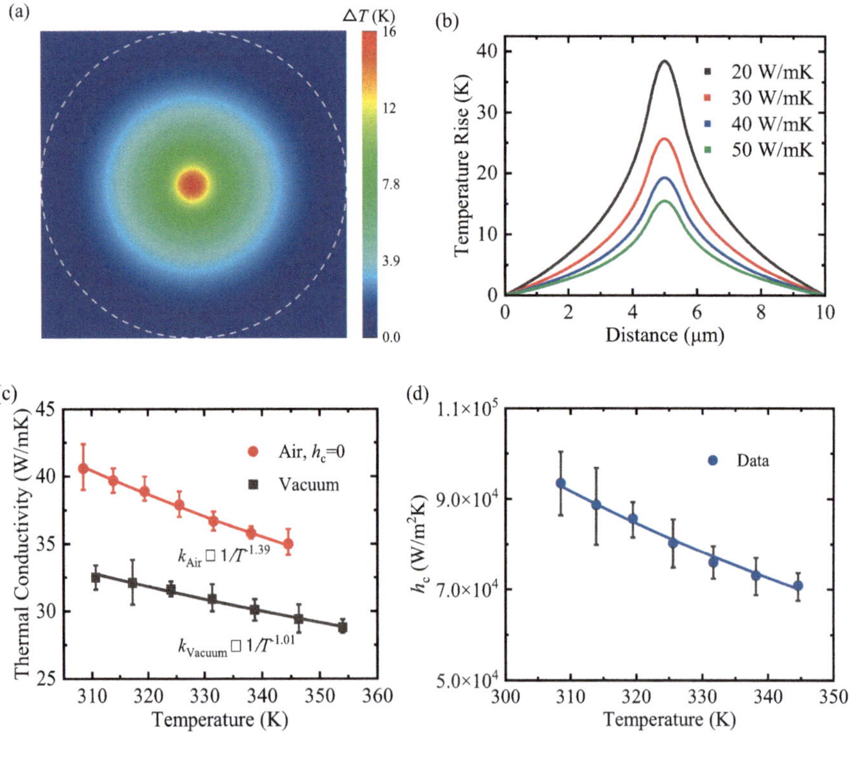

Fig. 4.4 (**a**) Calculated spatial distribution of temperature rise in the MoS_2 flake. (**b**) Temperature rise profiles of the MoS_2 flake at different thermal conductivity values. (**c**) Thermal conductivity of the MoS_2 flake as a function of temperature in air and vacuum. (**d**) Measured h_c as a function of temperature in air. (Reprinted with permission from American Institute of Physics ref. [4])

To evaluate the influence of air molecules on heat transfer, the total thermal conductivity (k_{total}) of the MoS_2 flake is measured in air. The parameter k_{total} corresponds to the effective thermal conductivity of the integrated air-MoS_2 system [15]. Figure 4.4c shows how the total thermal conductivity varies with temperature in air. k_{total} ranges from 40.6 to 35.0 W/mK over a temperature range of 308–345 K. These values are approximately 24.9% higher than the thermal conductivity measured under vacuum conditions. This result clearly indicates that heat transfer to the surrounding air significantly influences the overall thermal behavior of the MoS_2 flake. Moreover, k_{total} decreases more steeply with temperature (following $k_{total} \propto T^{-1.39}$) than that in vacuum. The main reason for the sharper decline of k_{total} at high temperatures is less efficient heat conduction between air molecules and the MoS_2 surface [16].

To evaluate heat conductance between air molecules and the MoS_2 flake, additional thermal dissipation pathways are included in the 3D simulation. The k of the MoS_2 flake is set to the value obtained under vacuum, and radiation effects are disregarded. A h_c is used to represent the conduction of heat between air molecules and the MoS_2 flake. This coefficient is calibrated at each laser power level so that the simulated temperature matches the experimental data. As shown in Fig. 4.4d, the values of h_c (which are influenced by uncertainty in k) range from 9.35×10^4 to 7.09×10^4 W/m^{2}K over the temperature interval of 308–345 K. The observed decrease in h_c is mainly due to temperature-induced changes in the thermal exchange between air molecules and the MoS_2 surface, which affect the density and kinetic energy of the surrounding air [17]. The h_c values obtained in this experiment are substantially higher than those commonly reported for natural convection (typically 0–10 W/m^{2}K) [18]. At the micro/nanoscale, viscous forces exceed buoyancy forces, suppressing natural convection and making heat conduction the dominant mechanism of thermal transfer between air and a heated surface [19]. These results highlight the considerable influence of air molecules on heat transport in 2D materials, especially under low-pressure conditions where the models of convective heat transfer may no longer apply.

4.4 Heat Convection at Nanoscale Hotspots

While the steady-state laser Raman measurements provide spatially resolved insights into convective heat transfer in 2D structures, they remain limited to micron-scale resolution. To investigate heat transfer at the nanoscale, a localized-heating Raman approach is employed, which combines plasmonic hotspot generation with high-resolution temperature mapping. This approach allows direct characterization of heat transfer at spatial scales inaccessible to conventional Raman mapping, providing a powerful tool for understanding nanoscale energy transport.

In this work, a tip-patterned Au nanorod array is fabricated and is irradiated with a Raman laser to generate nanoscale hotspots, and Raman spectra is captured and is used to directly measure both interfacial thermal conductance and convective heat transfer coefficients at these hotspots. Experimental results are complemented by numerical simulations to accurately resolve the localized heat dissipation. Template-assisted lithography is employed to create gold nanopatterns on a silicon substrate. An ultrathin alumina membrane (UTAM) serves as the template, while polymethyl methacrylate (PMMA) is used as a supporting material. After coating the template with a 200 nm gold layer and removing the UTAM, an array of gold nanorods is formed on the masked substrateA chemical vapor deposition (CVD)-grown single-layer graphene sheet measuring 1×1 cm^2 is then transferred onto the gold nanorods of 330 nm diameter (Au330/Si) substrate. Unlike the full contact achieved between graphene and the flat gold film/silicon substrate (AuF/Si), the graphene contacts the Au330/Si surface only at the top surfaces of the gold nanorods. A second sample with gold nanorods of 240 nm diameter (Au240/Si) is also prepared using the same method.

The temperature rise of graphene on both AuF/Si and Au330/Si is measured using the same Raman spectrometer. As shown in Fig. 4.5a and b, the G-band peak shifts to a lower wavenumber for both Gr/AuF/Si and Gr/Au330/Si due to laser-induced heating. However, under the same laser power, the amount of peak shift differs between the two samples. The average temperature increase per unit laser power is 1.38 ± 0.62 K/mW for Gr/AuF/Si and 3.52 ± 1.01 K/mW for Gr/Au330/Si. This difference may be attributed to enhanced optical fields caused by the nanorods, as well as distinct heat transfer mechanisms at the interface.

When exposed to laser light, gold nanostructures generate localized surface plasmon resonance (LSPR), which enhances the absorption of laser energy in the graphene layer placed on top of the nanorods. In addition to the LSPR effect, the structure of the substrate also significantly influences heat transfer: only certain areas of the graphene are in contact with the Au nanorods (Au330), while the remaining sections are suspended and exposed to air on both sides, allowing heat to be dissipated through convection and radiation.

To analyze convective heat transfer, it is necessary to evaluate thermal conduction at the graphene-gold interface in advance. Using the measured temperature rise of graphene on AuF/Si, the interfacial thermal conductance can be determined firstly based on finite element (FE) simulations (see Fig. 4.6a). For the Au330/Si

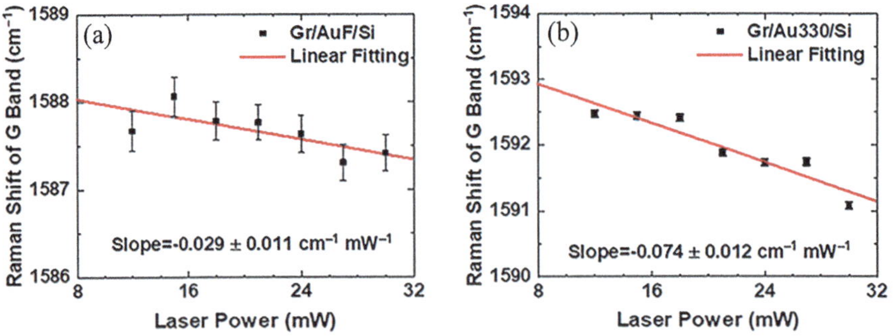

Fig. 4.5 Linear fitting of the *G-band* peak wavenumber against the laser power for (**a**) Gr/AuF/Si and (**b**) Gr/Au330/Si in the heating range from 12 to 30 mW [20]

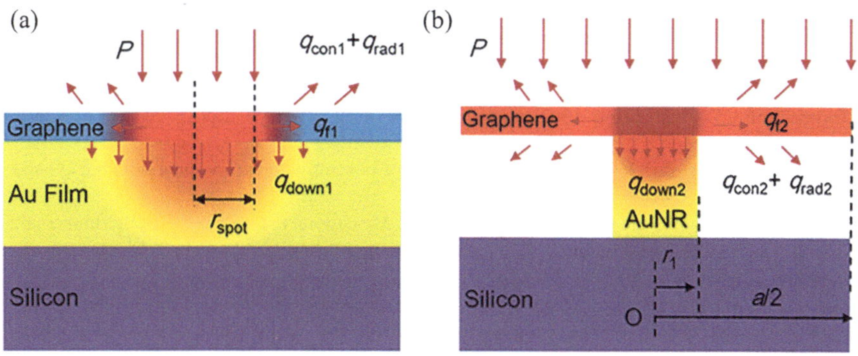

Fig. 4.6 Two heat transfer mechanisms observed in (**a**) Gr/AuF/Si and (**b**) Gr/Au330/Si during laser heating [20]

substrate, the top surface and suspended bottom region of graphene are influenced by air convection (see Fig. 4.6b). Radiative heat loss accounts for only 0.002%–0.003% of the total dissipation and is neglected in the calculations. By matching the simulated temperature rise in the laser spot to the experimental value, the h values around the graphene nanoscale hotspot on the gold nanopattern is determined to be 1928^{+155}_{-147} W/m^{2}K^1. Convective heat loss accounts for approximately 26.6% of the total thermal dissipation in Gr/Au330/Si. The high convection effect, combined with low interfacial thermal conductance, limits heat conduction into the Au nanorod substrate.

4.5 Summary

This chapter investigates convective heat transfer in low-dimensional materials ranging from microwires to nanoscale hotspots. Steady-state Raman mapping enable simultaneous characterization of thermal conductivity and convective heat transfer coefficients, revealing the significant contribution of air-mediated energy dissipation at small scales. The method extends to 2D materials with suspended structures and large surface exposure that amplify convective effects. Localized-heating Raman thermometry with plasmonic nanostructures enables probing of interfacial thermal conductance and convective heat transfer at the nanoscale.

References

1. Jia, C., Lin, Z., Huang, Y., Duan, X.: Nanowire electronics: from nanoscale to macroscale. Chem. Rev. **119**(15), 9074–9135 (2019)
2. Hunter, N., Azam, N., Zobeiri, H., Wang, R., Mahjouri-Samani, M., Wang, X.: Interfacial thermal conductance between monolayer WSe2 and SiO2 under consideration of radiative electron–hole recombination. ACS Appl. Mater. Interfaces. **12**(45), 51069–51081 (2020)

3. Gao, J., Xie, D., Xiong, Y., Yue, Y.: Thermal characterization of microscale heat convection in rare-gas environment by a steady-state "hot wire" method. Applied Physics Express. **11**(6), 066601 (2018)
4. Li, D., Huang, X., Huang, D., Zhang, N., Liao, D., Zhang, Z., Wu, S., Yue, Y.: Thermal transport and accommodation coefficient at the air–MoS2 interface. Appl. Phys. Lett. **126**(19), 126 (2025)
5. Yue, Y., Huang, X., Wang, X.: Thermal transport in multiwall carbon nanotube buckypapers. Phys. Lett. A. **374**(40), 4144–4151 (2010)
6. Zhao, D., Qian, X., Gu, X., Jajja, S.A., Yang, R.: Measurement techniques for thermal conductivity and interfacial thermal conductance of bulk and thin film materials. J. Electron. Packag. **138**(4), 040802 (2016)
7. Dai, W., Wang, Y., Li, M., Chen, L., Yan, Q., Yu, J., Jiang, N., Lin, C.T.: 2D materials-based thermal interface materials: structure, properties, and applications. Adv. Mater. **36**(37), 2311335 (2024)
8. Li, M., Li, C., Wang, J., Xiao, X., Yue, Y.: Parallel measurement of conductive and convective thermal transport of micro/nanowires based on Raman mapping. Appl. Phys. Lett. **106**(25), 253108 (2015)
9. Peirs, J., Reynaerts, D., Van Brussel, H.: In: Scale effects and thermal considerations for micro-actuators, Proceedings. 1998 IEEE International Conference on Robotics and Automation (Cat. No. 98CH36146), IEEE. 1516–1521 (1988)
10. Sokół, A.K., Sarzała, R.P.: Thermal management of GaInNAs/GaAs VECSELs. Opto-Electron. Rev. **21**(2), 191–198 (2013)
11. Giberti, A., Guidi, V., Vincenzi, D.: A study of heat distribution and dissipation in a micromachined chemoresistive gas sensor. Sensors Actuators B Chem. **153**(2), 409–414 (2011)
12. Guan, N., Liu, Z., Zhang, C., Jiang, G.: Natural convection heat transfer on surfaces of copper micro-wires. Heat Mass Transf. **50**(2), 275–284 (2014)
13. Wang, Z., Tang, D.: Investigation of heat transfer around microwire in air environment using 3ω method. Int. J. Therm. Sci. **64**, 145–151 (2013)
14. Chen, J., Gui, X., Wang, Z., Li, Z., Xiang, R., Wang, K., Wu, D., Xia, X., Zhou, Y., Wang, Q.: Superlow thermal conductivity 3D carbon nanotube network for thermoelectric applications. ACS Appl. Mater. Interfaces. **4**(1), 81–86 (2012)
15. Klaas, S.T.M.J.V., Tielrooij, J., Reig, D.S., Varghese, S., Farris, R., Block, A., Mehew, J., Sledzinska, M., El Sachat, A., Chávez-Ángel, E.: Unraveling heat transport and dissipation in suspended MoSe2 from bulk to monolayer. Adv. Mater. **34**(10), 2108352 (2022)
16. Pop, E., Mann, D., Wang, Q., Goodson, K., Dai, H.: Thermal conductance of an individual single-wall carbon nanotube above room temperature. Nano Lett. **6**(1), 96–100 (2006)
17. Alam, M., Raghu, A., Haque, M.A., Muratore, C., Voevodin, A.A.: Structural size and temperature dependence of solid to air heat transfer. Int. J. Therm. Sci. **73**, 1–7 (2013)
18. Bergman, T.L.: Fundamentals of Heat and Mass Transfer. John Wiley & Sons (2011)
19. Gao, J., Xie, D., Cai, K., Chen, Y., Yue, Y.: Finite differential hot-wire method for localized heat transfer characterization at microscale. Appl. Phys. Express. **13**(10), 105003 (2020)
20. Xu, S., Xu, Y., Zhang, J., Gao, J., Wang, X., Zhang, X., Yue, Y.: Ballistic transport enhanced heat convection at nanoscale hotspots. J. Appl. Phys. **136**(16), 164306 (2024)

Chapter 5
Raman Characterization of Interfacial Thermal Transport

Abstract Energy dissipation in nanoelectronics has emerged as a critical bottleneck that limits both storage density and computational speed. Two-dimensional (2D) materials such as graphene and MoS_2, with their atomic-layer thickness and exceptional physical properties, offer promising solutions to these challenges. However, their ultrathin nature requires substrate support for mechanical stability, which in turn introduces strong interfacial effects that dominate overall heat transport. A significant portion of the dissipated energy transfers through the 2D material–substrate interface, making interfacial thermal resistance a key determinant of device thermal performance (Yue et al., Small. 7:3324–3333, 2011). This chapter focuses on Raman-based characterization techniques for investigating interfacial thermal transport. Section 5.1 introduces the fundamental principles of Raman thermometry as applied to interface studies. Section 5.2 presents the Raman penetration method. Section 5.3 describes the two-step Raman method for decoupling in-plane and cross-plane heat transport contributions. Section 5.4 discusses thermal resistance at nanoscale contacts, addressing localized phonon scattering and imperfect bonding effects. Section 5.5 explores strategies for modulating interfacial thermal resistance through interface engineering, surface functionalization, and heterostructure design.

Keywords Interfacial thermal transport · Raman thermometry · 2D materials · Thermal boundary resistance

Y. Yue, *Raman Thermometry*, https://doi.org/10.1007/978-981-95-5985-5_5

5.1 Principle of Raman Thermometry on Interface

Thermal transport across atomic-layer interfaces is complex due to limited spatial dimensions and uncertainties in the interfacial interactions between adjacent materials. In the case of atomic-layer materials, the thermal characterization becomes increasingly challenging as the sample thickness decreases. First, samples with atomic-thin thickness exhibit extremely fast heat dissipation. As a result, traditional electrical methods cannot meet the requirements due to the limitations of the instrumentation. In optical measurements such as the ultrafast pump-probe method, defining the surface temperature is challenging because a significant portion of the incident laser penetrates the material. Therefore, there is a need to develop a new method that can directly distinguish the surface temperature of the sample from that of the substrate, even at such small scales, without any additional treatment.

Raman thermometry relies on the fact that phonon frequencies influenced by anharmonic interatomic potentials are temperature dependent. In commonly used contact-mode temperature probing methods, such as thermocouples and thermal resistors, a probe must be attached to the sample's surface, with temperature measured based on the voltage across the resistors. Thermal paste is often applied to improve contact between the probe and the surface; however, it may introduce contamination or damage to the sample. Non-contact optical temperature measurements on are preferable and offer broader applicability. Raman characterization is particularly effective for measuring thermal transport across an ultrathin film/substrate interface. It is because that the Raman excitation laser can penetrate the film to reach the substrate, providing simultaneous temperature data for both the film and the substrate, thereby enabling the measurement of the temperature differential across the interface. As shown in Fig. 5.1, Raman thermometry allows for the direct and simultaneous determination of temperature information for the MoS_2 film and c-Si substrate, which can then be used to calculate the interface thermal resistance.

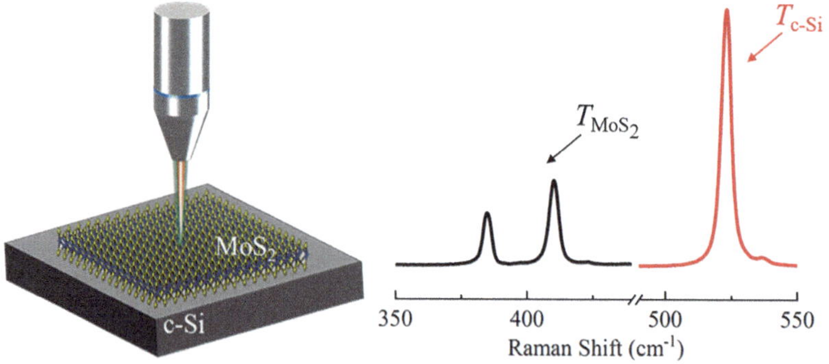

Fig. 5.1 Schematic of experimental setup for Raman-based characterization of MoS_2/c-Si interface. (Reprinted from ref. [1], Copyright(2023), with permission from American Institute of Physics)

5.2 Raman Penetration Method

Graphene has garnered significant attention due to its unique properties, including electric-field effects, the quantum Hall effect, and high charge-carrier mobility. In this section, a graphene on a SiC substrate is used to explore the feasibility of Raman method for exploring the energy dissipation across the interface and measuring the local thermal contact resistance. To characterize the thermal contact resistance between graphene and SiC, the temperature difference between them must first be determined. As shown in Fig. 5.2, the graphene sheet is connected to two electrodes. An electrical current is passed through the graphene to induce steady-state Joule heating. At the same time, a confocal micro-Raman experiment is conducted by irradiating the central area of the graphene layer with an excitation laser.

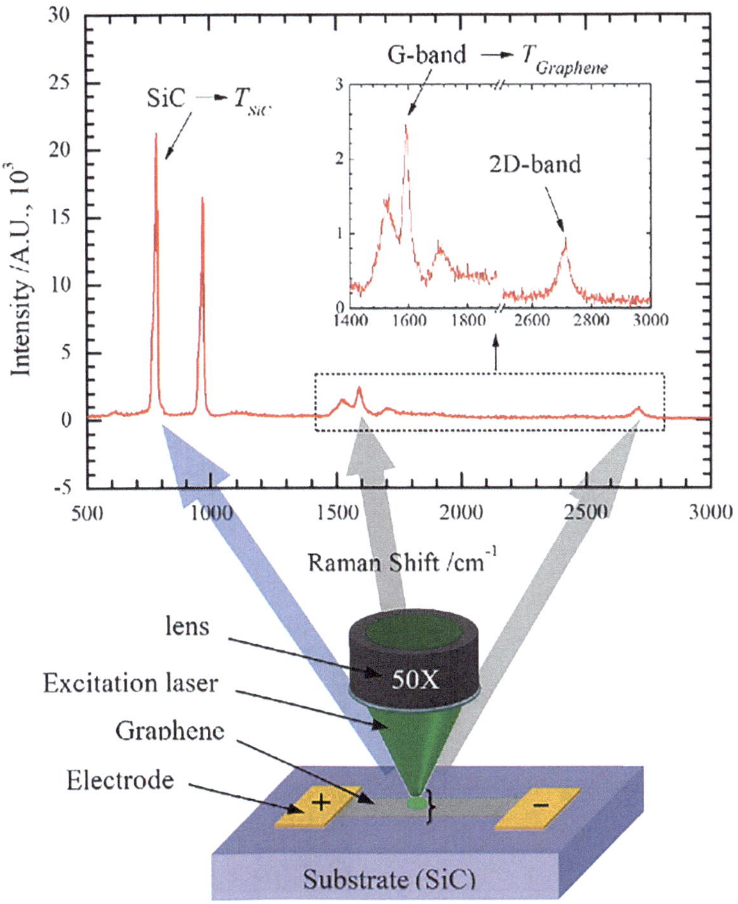

Fig. 5.2 The schematic of the electrical heating Raman-probing method to measure interfacial thermal contact resistance between graphene and 4H-SiC. (Reprinted from ref. [2], Copyright(2011), with permission from Wiley)

Based on the measured Raman signal, the temperature of the graphene sheet and the SiC substrate can be distinguished simultaneously with high spatial resolution in the vertical direction. The interfacial thermal resistance can be readily determined from $R_{tc} = (T_{graphene} - T_{SiC})A/(I^2R)$, where A is the graphene area, R is its electrical resistance, I is the applied current, $T_{graphene}$ and T_{SiC} are the temperature rise of graphene and SiC, respectively. The heat generated from Joule heating is dissipated in two directions: one part crosses the plane to the interface and the other part dissipates along the graphene layer. In the joule-heating experiment, the second part is negligible because the heating is uniform and almost no temperature gradient exists in the in-plane direction.

As mentioned above, the thermal resistance at the graphene/SiC interface is expressed as $R_{tc} = (T_{graphene} - T_{SiC})A/(I^2R)$. To obtain a sensible temperature difference between the graphene and the SiC, a controllable heating energy is applied on the graphene flake. The results for the relationship between temperature and input power are illustrated in Fig. 5.3. To improve the accuracy of the measurement, a linear fitting of the relationships between temperature and power was conducted. The thermal resistance is calculated by combining the slope difference of the fittings and the sample area, as shown in Fig. 5.3. R_{tc} is calculated as $5.30^{+0.46}_{-0.46} \times 10^{-5}$ Km²/W, where the uncertainty comes from the linear fit of the temperature difference. The measured interfacial thermal resistance is significantly higher than the values reported for graphene on other substrate materials [3]. The contact resistance between graphene and silicon dioxide, measured using the 3ω method, ranges from 5.6×10^{-9} to 1.2×10^{-8} Km²/W. Lyeo et al. [4] measured the thermal-contact

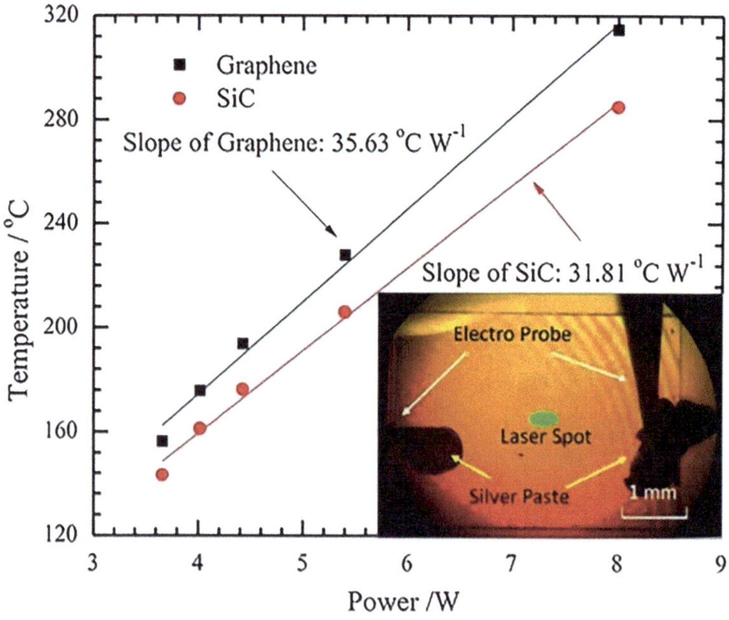

Fig. 5.3 The relationship between temperature and heating power. (Reprinted from ref. [2], Copyright(2011), with permission from Wiley)

resistance of interfaces between highly dissimilar materials to be approximately 3.0×10^{-8} Km^2/W. The higher thermal contact resistance measured here results from the thermal expansion mismatch between graphene and SiC under Joule heating. This mismatch causes interface delamination and separation, significantly enhancing local phonon scattering. The thermal probing technique presented in this section offers a promising approach for achieving nanometer resolution in temperature measurements.

5.3 Two-step Raman Method

Most studies on thermal transport in graphene/substrate structures have primarily examined either in-plane thermal conduction or thermal conductance across the interface. However, this process is inherently coupled: energy dissipation at the interface and in-plane thermal transport occur simultaneously, producing a localized hot spot. Therefore, a comprehensive understanding of thermal transport in graphene–substrate systems require simultaneous consideration of both transport pathways. To address this issue, an in-situ two-step Raman technique has been established to simultaneously quantify the interfacial thermal conductance between graphene and substrate, as well as the in-plane thermal conductivity of this supported graphene. This two-step Raman method consists of a joule-heating step to characterize interfacial thermal conductance, and a laser-heating step to measure in-plane thermal conductivity.

As shown in Fig. 5.4, the electrical heating method is the first step, where steady-state joule heating is induced in graphene by a constant current. Heat in the graphene layer is conducted through the interface to the substrate, resulting in a

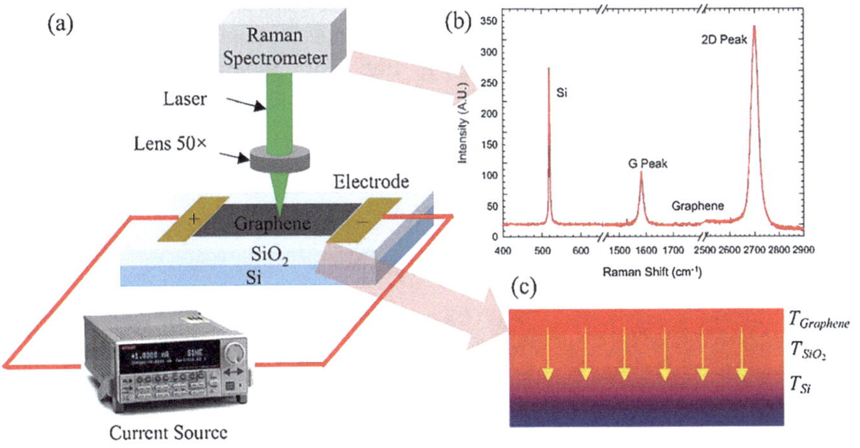

Fig. 5.4 (**a**) Schematic of the Raman thermometry technique for characterizing interfacial thermal resistance between graphene and silicon. (**b**) Raman spectrum of graphene on a silicon substrate and (**c**) heat transfer from graphene to the substrate. (Reprinted from ref. [5], Copyright(2017), with permission from Elsevier)

temperature gradient across the interface. This temperature difference only depends on the heating density and the thermal resistance across the interface. As a single atomic layer, graphene captures about 2.3% of the laser energy while reflecting under 0.1%, with most of the laser penetrating graphene and reaching the substrate [6]. This results in two distinct Raman peaks on the spectrum, corresponding to both graphene and substrate. The temperature of the surrounding materials can be determined from these spectral peaks and subsequently used to assess the interfacial thermal resistance.

In contrast to the first step, where laser energy is minimized, the laser in the second step requires enough intensity to induce a substantial temperature rise. This temperature rise is influenced by both the interfacial thermal conductance and in-plane thermal conductivity (Fig. 5.5a). Thermal transport in graphene can be described as follows [7]

$$q = 2\pi k r_1 \delta \theta m \frac{K_1(mr_1)I_1(mr_2) - K_1(mr_2)I_1(mr_1)}{K_0(mr_1)I_1(mr_2) + K_1(mr_2)I_0(mr_1)} \tag{5.1}$$

where q denotes the heat flux across boundary of the heated portion of graphene, k is the in-plane thermal conductivity of the supported graphene, d represents the thickness of graphene, θ is the temperature difference at boundary of the heated section, $m = \sqrt{2h / k\delta}$, h is the interfacial thermal conductance between graphene and SiO_2, r_1 corresponds to the radius of heated region and r_2 to that of the unheated region, I_0, K_0, I_1, K_1 refer to the zeroth- and first-order modified Bessel functions of the first and second kinds [7]. In Eq. (5.1), q is obtained by subtracting the interfacial thermal transport measured in the first step. Thus, the in-plane thermal conductivity of graphene is the only unknown parameter to be calculated in this equation.

Noted that only a limited amount of laser energy is absorbed by graphene during the measurement. For a 532 nm laser, the absorptivity of monolayer graphene is theoretically calculated to be only 2.3% [6]. Even a small difference in laser absorption can lead to a significant variation in in-plane thermal conductivity. For the graphene/SiO_2/Si interface illustrated in Fig. 5.5b, multiple transmissions and reflections of the laser occur. To minimize measurement uncertainty, Eq. (5.1) is formulated as:

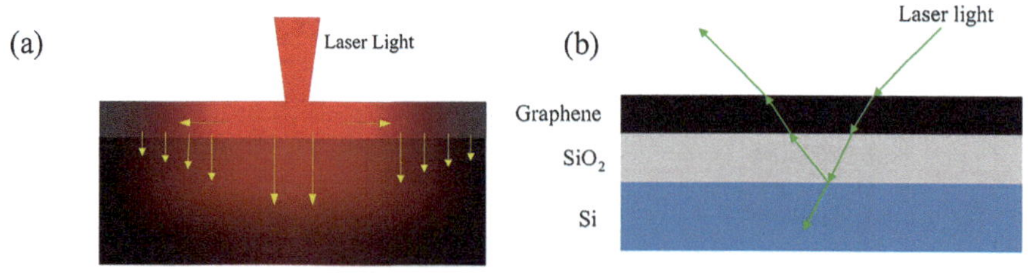

Fig. 5.5 (**a**) Heat transfer under localized laser heating. (**b**) Laser reflection and refraction in the presence of a SiO_2 layer. (Reprinted from ref. [5], Copyright(2017), with permission from Elsevier)

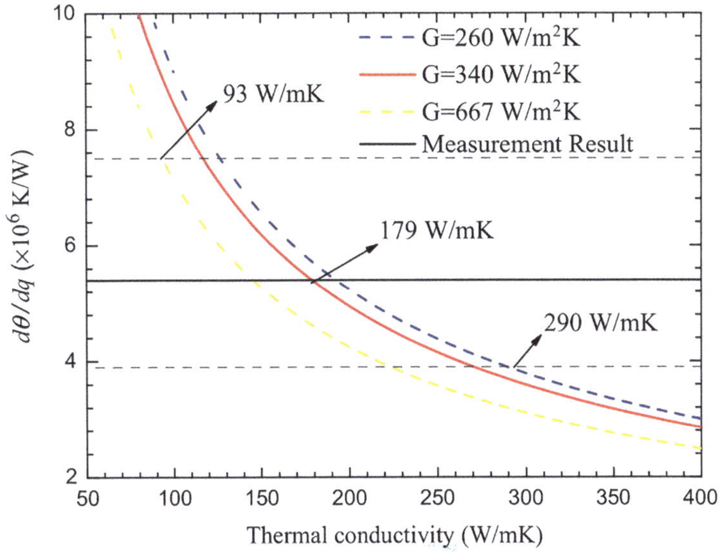

Fig. 5.6 Theoretical thermal conductivity curve of supported graphene. (Reprinted with permission from ref. [5])

$$\frac{d\theta}{dq} = \frac{1}{2\pi k r_1 \delta m} \times \frac{K_0(mr_1)I_1(mr_2) + K_1(mr_2)I_0(mr_1)}{K_1(mr_1)I_1(mr_2) - K_1(mr_2)I_1(mr_1)} \tag{5.2}$$

The interfacial thermal resistance can be derived by considering different powers applied during the experiment. Interfacial thermal resistance is characterized as $R_{g/SiO_2} = 2.94^{+1.9}_{-1.44} \times 10^{-3}$ m²K/W, and the interfacial thermal conductance is obtained as $G_{g/SiO_2} = 340^{+327}_{-80}$ W/mK accordingly. By measuring the temperature rise of graphene at different laser powers, the in-plane thermal conductivity can be calculated. Based on the geometry structure, the measured interfacial thermal conductance, and the temperature rise at different laser-heating powers, a theoretical thermal conductivity curve for supported graphene is obtained as shown in Fig. 5.6. The graphene exhibits a thermal conductivity of 179^{+111}_{-86} W/mK. This two-step Raman technique enables assessment of both in-plane and cross-plane thermal transport in 2D interface materials, which is essential for a comprehensive understanding of energy dissipation across these interfaces.

5.4 Thermal Resistance at the Nanoscale Contact

The thermal resistance of nanoscale point contacts has become a critical issue in thermal management. At small scales, phonon transport deviates significantly from classical predictions, with ballistic effects dominating over diffusive conduction.

This breakdown of Fourier's law complicates both theoretical modelling and experimental characterization, making the accurate determination of thermal resistance at nanoscale point contacts particularly challenging. Conventional techniques lack the spatial resolution required to resolve the localized thermal behaviour of individual nanoscale contacts. In this regard, Raman thermometry, owing to its high spectral sensitivity and spatial resolution, provides a powerful approach for probing phonon transport at nanoscale contacts.

In this section, studies on heat transfer at the nanotip-graphene interface are presented to investigate the thermal resistance of nanoscale point contacts. Au nanorod tips induce a near-field effect, known as local surface plasmon resonance (LSPR), which enhances the absorption of incident energy by the graphene layer. This process generates LSPR-induced nanoscale hotspots within the graphene layer (Fig. 5.7a). The steady-state heat transfer within the sample is illustrated in Fig. 5.7b. Part of the heat is conducted laterally along the graphene plane toward the heat sink, while the remainder is transferred perpendicularly through the interface between graphene and the Au nanopillars to the substrate.

Within the laser-irradiated region, the graphene temperature represents that of a high-temperature heat source (hot side). Since the temperature rise in the silicon substrate is negligible (as confirmed by experimental measurements), the ambient and substrate temperature is considered the heat sink temperature (cold side). The in-plane and cross-plane thermal resistances of graphene are connected in parallel. The cross-plane resistance consists of the interfacial thermal resistance between graphene and Au nanopillar array, the intrinsic thermal resistance of Au nanopillars, and the interfacial thermal resistance between Au nanopillars and silicon substrate. Based on the equivalent thermal resistance network, once the high-temperature

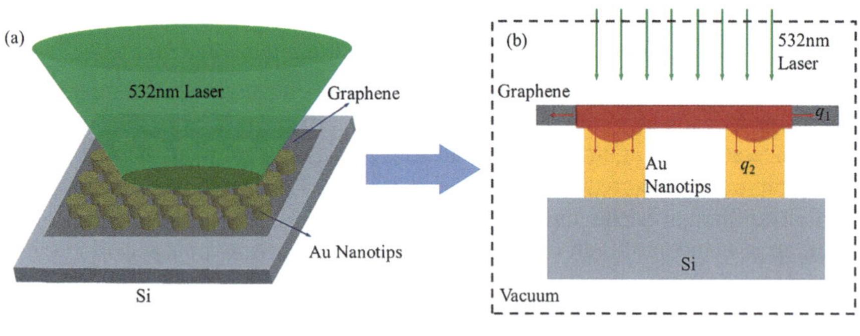

Fig. 5.7 Thermal transport model in nanopillars supported graphene. (**a**) Overview of the laser irradiation on the nanopillar supported graphene layer; (**b**) Laser heating and thermal transport from graphene layer to the substrate through nanopillars [8]

source is determined, the total thermal resistance can be calculated, allowing further analysis to extract interfacial thermal resistance between graphene and Au nanopillar.

The temperature rise of graphene on Au nanopillar array is determined using the calibrated Raman shift-temperature coefficient. As the laser power increases from 2 mW to 60 mW, the Raman shift of the characteristic silicon peak (521 cm^{-1}) exhibits minimal variation. During the laser heating process, the silicon substrate can be assumed to experience a negligible temperature rise. Under the same heating conditions, the graphene *G* peak on the Au nanopillar array exhibits a fitted Raman shift-power coefficient of −0.089 cm^{-1}/mW. Taking the case of 45 mW laser power as an example, the average surface temperature of graphene in vacuum reaches (146.8 ± 11.79°C), while the temperature rise at the surface of the silicon substrate is only about 4 °C.

The experiment is conducted under vacuum conditions, where heat convection can be neglected. In addition, the radiative heat exchange between graphene and the surrounding environment is minimal and thus disregarded. The governing equation for steady-state heat conduction in graphene is expressed as:

$$\frac{1}{r}\frac{\mathrm{d}}{\mathrm{d}r}\left(r\frac{\mathrm{d}T}{\mathrm{d}r}\right)-\frac{G_{\mathrm{Gr/Au}}}{k_{\mathrm{Gr}}t_{\mathrm{Gr}}}\left(T-T_{\mathrm{a}}\right)+\frac{\dot{q}_{\mathrm{Gr2}}}{k_{\mathrm{Gr}}}=0 \tag{5.3}$$

where $G_{\mathrm{Gr/Au}}$ is the interfacial thermal conductance between graphene and the Au nanopillar array (units of W/m^2K), which is the key parameter to be determined. k_{Gr} and t_{Gr} denote the in-plane thermal conductivity and thickness of graphene, respectively. q_{Gr2} represents the volumetric heat generation rate in graphene under laser irradiation, expressed in cylindrical coordinates as: $q_{\mathrm{Gr2}} = I\alpha_{\mathrm{Gr}}M / t_{\mathrm{Gr}}$, where $\alpha_{\mathrm{Gr}} = 0.023$ is the optical absorption coefficient of monolayer graphene, $M = 3.2$ is the electric field intensity enhancement factor. Noted that graphene is only partially in contact with the Au nanopillars within the laser-irradiated region, with an estimated contact area ratio of approximately 51%. The correlation between the effective interfacial thermal resistance and temperature rise (146.8 ± 11.79 °C) is established through finite element method simulations, as illustrated in Fig. 5.8a. Based on this analysis, the thermal contact resistance between the Au nanotip and graphene is determined to be 4.6×10^{-5} (m^2K)/W. This work is just a demonstration that Raman thermometry can be employed to quantify interfacial heat transfer at a nanoscale point contact formed between inverted probes and 2D materials. Such measurements provide valuable insights for the design and optimization of point-contact configurations in device thermal management. In addition, there are significant implications for advancing the understanding and application of scanning probe microscopy and nanolithography techniques, where localized heating and thermal transport play critical roles.

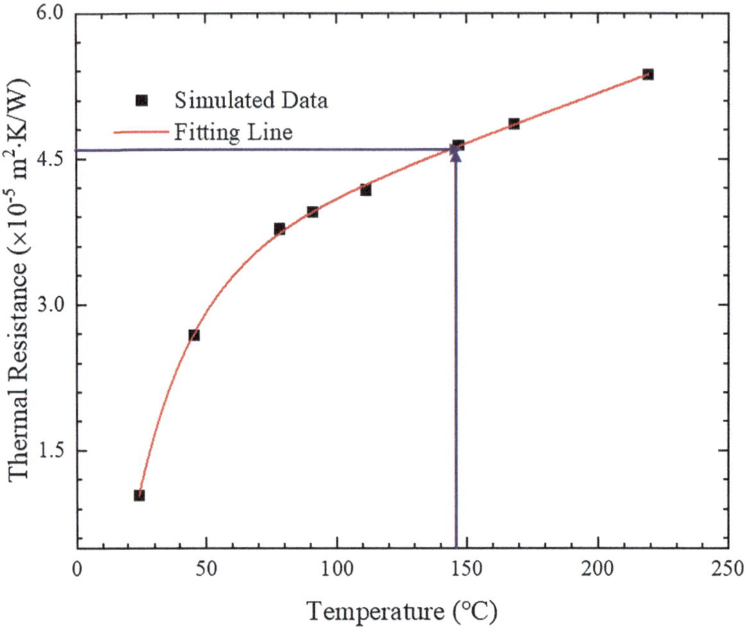

Fig. 5.8 Determination of interfacial thermal resistance between Au nanotip and graphene. (Reprinted from ref. [8], Copyright(2024), with permission from Pleiades Publishing)

5.5 Modulation of Interfacial Thermal Resistance

Theoretical studies reveal that interfacial thermal transport is highly sensitive on the interface interactions, which are typically described as the Van der Waals (vdW) force in molecular dynamics simulations [9]. Notably, by adjusting the interlayer coupling factor, the interfacial thermal resistance can be reduced by more than two orders of magnitude [10]. Another practical strategy to improve interfacial thermal transport is surface modification, such as introducing functional groups or forming chemical bonds at the interface. However, above mentioned approaches are either predominantly based on calculations or difficult to implement in practical applications. Therefore, further research is necessary to identify a practical and easily achievable approach. In this section, the enhancement of interfacial thermal transport is introduced, focusing on repeated annealing of the graphene/Si/SiO_2 interface.

The schematic diagram of thermal transport between graphene and Si substrate is presented in Fig. 5.9. The measured total thermal resistance R includes contributions from the graphene-silicon dioxide interface $R_{G\text{-}SiO_2}$, silicon dioxide layer R_{SiO_2}, and silicon dioxide-silicon interface $R_{Si\text{-}SiO_2}$. In the experiment, the sample is heated by a laser beam focused through a lens. The focused beam generates a highly localized heating region, and then the heat flux flows from the laser spot to the substrate. Three heat transfer channels, including heat conduction along the in-plane

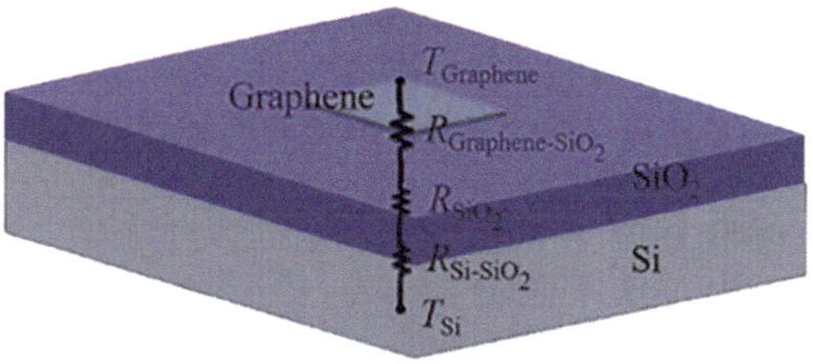

Fig. 5.9 Schematic of the thermal circuit network for split *R* of Si and SiO_2 from the total *R*. (Reprinted from ref. [11], Copyright(2023), with permission from Elsevier)

direction of graphene, heat transfer across the interface between the substrate and graphene, and air convection, are responsible for transferring heat to achieve the final thermal equilibrate state.

Accordingly, the heat conduction equation of the model can be expressed as:

$$k_G t_G \frac{1}{r}\frac{\partial}{\partial r}\left(r\frac{\partial T_G}{\partial r}\right) - \frac{T_G(r) - T_{sub}}{R} - h\left(T_g - T_{air}\right) + \alpha_{eff} q(r) = 0 \tag{5.4}$$

where k_G, t_G, and R are the thermal conductivity, thickness and interfacial thermal resistance of graphene, respectively. T_G and T_{sub} represent the temperature of the graphene and substrate. The first term of this equation describes the heat transport in graphene using a cylindrical coordinate system. To simplify the model, a one-dimensional approximation is adopted, considering the atomic-scale thickness of graphene ($t_G = 0.335$ nm) and the power distribution of the laser beam. The second term $(T_G(r) - T_{sub})/R$ represents cross-plane heat transfer across the graphene-substrate interface. The third term $h(T_g - T_{air})$ accounts for convective heat loss, which can be neglected due to the relatively small temperature rise of graphene. The final term represents the laser-induced heat source, where $\alpha_{eff} = 2.3\%$ is the effective absorption factor of graphene [12] and q(r) indicates the Gaussian spatial distribution of the laser energy flux, expressed as:

$$\mathrm{q}(r) = q_0 \exp\left(-\frac{2r^2}{r_0^2}\right) \tag{5.5}$$

where $q_0 = P/\pi r_0^2$ is the laser power density, and r_0 is the laser spot radius. 3D numerical modelling can be conducted, and the presupposed k_G are used to solve the model and obtain the result of R [13].

The interfacial thermal resistance R extracted from the model and Raman-measured temperature is obtained 9.51×10^{-5} m^{2}K/W, which is three orders of magnitude higher than previously reported calculated results. This discrepancy is primarily attributed to the strength of the interface coupling. Recent work by Tang et al. [14] has demonstrated that interfacial thermal resistance can be altered by up to five orders of magnitude depending on the contact between the graphene and the substrate. The Si/SiO_2-supported graphene is annealed and Raman measured sequentially to obtain the R value corresponding to the different number of annealing (Fig. 5.10a). With the increase of annealing times, the values of R slumped over an order of magnitude, from 9.5×10^{-5} m^{2}K/W to 5.0×10^{-6} m^{2}K/W. It should be mentioned that k_G used to determine the value of R is also strongly influenced by annealing and drops sharply after annealing [15]. The effect of presupposed k_G is considered here, and it is found that changes of k_G do not have a significant influence on acquiring the final value of R.

The enhancement of interfacial thermal transport induced by annealing is complex and multifaceted. A major contributing factor is the presence of wrinkles, which are difficult to avoid or eliminate during the growth and transfer process, particularly in large-scale graphene synthesis [16]. As shown in Fig. 5.10b, wrinkles are still there after 1 h of annealing at 600 K, but compared with the surface image of graphene before annealing, the surface wrinkles undergo a flattening process. The reduction in RMS surface roughness confirms that thermal treatment can effectively smooth the graphene surface. Another factor contributing to the enhancement of interfacial thermal transport is the presence of a possible water layer [17]. During the transfer of CVD-grown monolayer graphene, wet processing can leave a thin water layer between the graphene and the substrate [18], which reduces the adhesion energy and weakens the graphene-substrate contact. Although graphene is hydrophobic, experiments have shown that water molecules can intercalate from its

Fig. 5.10 (**a**) Decreased interfacial thermal resistance with annealing treatment. (**b**) AFN images of graphene surface before and after annealing. (Reprinted from ref. [11], Copyright(2023), with permission from Elsevier)

edges due to the hydrophilicity of the substrate [19]. Annealing effectively removes this interfacial water layer, reducing the separation between graphene and the substrate and thereby enhancing thermal transport.

5.6 Summary

This chapter introduces Raman-based characterization of interfacial thermal transport. For steady-state Raman measurements, accurate thermal characterization requires precise evaluation and calibration of the sample's optical absorptivity. However, this process is often challenging and time-consuming, particularly for micro- and nanoscale materials. In addition, steady-state methods do not provide information on transient parameters, such as thermal diffusivity. To overcome the challenges associated with heating-level evaluation, calibration, and inevitable temperature deviations, a promising alternative for Raman-based thermal characterization is the use of a time-resolved approach. This method captures the transient thermal response of the material, enabling the determination of both thermal diffusivity and the normalized interfacial conductance $G/\rho c$ (ρ and c are the density and specific heat of the material, and G is the interfacial thermal conductance). The following chapter focuses on the principles and applications of transient Raman thermometry.

References

1. Li, D.S., Huang, X.N., Liu, Z.Y., Liu, W.X., Xu, S., Yue, Y.N.: Divergent interfacial thermal transport in MoS2/Si heterostructure over optical phonon modes. Appl. Phys. Lett. **123**(25), 252201 (2023)
2. Yue, Y.N., Zhang, J.C., Wang, X.W.: Micro/nanoscale spatial resolution temperature probing for the interfacial thermal characterization of epitaxial graphene on 4H-SiC. Small. **7**(23), 3324–3333 (2011)
3. Chen, Z., Jang, W., Bao, W., Lau, C.N., Dames, C.: Thermal contact resistance between graphene and silicon dioxide. Appl. Phys. Lett. **95**(16), 161910 (2009)
4. Lyeo, H.-K., Cahill, D.G.: Thermal conductance of interfaces between highly dissimilar materials. Phys. Rev. B. **73**(14), 144301 (2006)
5. Zhao, W., Chen, W., Yue, Y., Wu, S.J.: In-situ two-step Raman thermometry for thermal characterization of monolayer graphene interface material. Appl. Therm. Eng. **113**, 481–489 (2017)
6. Nair, R.R., Blake, P., Grigorenko, A.N., Novoselov, K.S., Booth, T.J., Stauber, T., Peres, N.M.R., Geim, A.K.: Fine structure constant defines visual transparency of graphene. Science. **320**(5881), 1308–1308 (2008)
7. Bergman, T.L.: Fundamentals of Heat and Mass Transfer. John Wiley & Sons (2011)
8. Xu, S., Deng, S., Huang, X., Huang, D., Yue, Y.: Study on heat transfer across nanotip contact interface under near-field effect. J. Eng. Thermophys. **45**(10), 3131–3136 (2024)
9. Ren, W.J., Ouyang, Y.L., Jiang, P.F., Yu, C.Q., He, J., Chen, J.: The impact of interlayer rotation on thermal transport across graphene/hexagonal boron nitride van der Waals heterostructure. Nano Lett. **21**(6), 2634–2641 (2021)

10. Zhang, L.N., Zhong, Y., Qian, X., Song, Q.C., Zhou, J.W., Li, L., Guo, L., Chen, G., Wang, E.N.: Toward optimal heat transfer of 2D-3D heterostructures van der Waals binding effects. ACS Appl. Mater. Interfaces. **13**(38), 46055–46064 (2021)
11. Liu, W., Sun, Q., Zhang, J., Huang, X., Yue, Y.: Annealing-induced one order of magnitude enhancement in heat dissipation at graphene/SiO2 interface. Mater. Today Nano. **24**, 100383 (2023)
12. Xu, Y.R., Jiang, J., Yang, B., Li, M.D., Xu, S., Yue, Y.N.: Comparison of thermal conductance of graphene/SiO2 and graphene/Au interfaces based on Raman optothermal method. Mater. Res. Express. **6**(11), 115603 (2019)
13. Yuan, P.Y., Wang, R.D., Wang, T.Y., Wang, X.W., Xie, Y.S.: Nonmonotonic thickness-dependence of in-plane thermal conductivity of few-layered MoS2: 2.4 to 37.8 nm. Phys. Chem. Chem. Phys. **20**(40), 25752–25761 (2018)
14. Tang, X.D., Xu, S., Zhang, J.C., Wang, X.W.: Five orders of magnitude reduction in energy coupling across corrugated graphene/substrate interfaces. ACS Appl. Mater. Interfaces. **6**(4), 2809–2818 (2014)
15. Kim, H.G., Kihm, K.D., Lee, W., Lim, G., Cheon, S., Lee, W., Pyun, K.R., Ko, S.H., Shin, S.: Effect of graphene-substrate conformity on the in-plane thermal conductivity of supported graphene. Carbon. **125**, 39–48 (2017)
16. Liu, N., Pan, Z.H., Fu, L., Zhang, C.H., Dai, B.Y., Liu, Z.F.: The origin of wrinkles on transferred graphene. Nano Res. **4**(10), 996–1004 (2011)
17. Peng, X.Y., Jiang, P.F., Ouyang, Y.L., Lu, S., Ren, W.J., Chen, J.: Reducing Kapitza resistance between graphene/water interface via interfacial superlattice structure. Nanotechnology. **33**(3), 035707 (2022)
18. Verguts, K., Schouteden, K., Wu, C.H., Peters, L., Vrancken, N., Wu, X.Y., Li, Z., Erkens, M., Porret, C., Huyghebaert, C., Van Haesendonck, C., De Gendt, S., Brems, S.: Controlling water intercalation is key to a direct graphene transfer. ACS Appl. Mater. Interfaces. **9**(42), 37484–37492 (2017)
19. Lee, D., Ahn, G., Ryu, S.: Two-dimensional water diffusion at a graphene-silica interface. J. Am. Chem. Soc. **136**(18), 6634–6642 (2014)

Chapter 6
Transient Raman Characterization of Materials

Abstract Conventional Raman thermometry for measuring thermal properties rely on steady-state approaches, such as Joule heating or two-laser configurations. While these methods have been widely applied, they often involve increased system complexity and considerable measurement uncertainty. Accurate implementation of steady-state Raman thermometry requires precise calibration of the relationship between temperature and Raman response, as well as reliable determination of laser absorption. Both factors are critical for quantitative accuracy but can be time-consuming, particularly when dealing with samples of varying optical properties. To overcome these limitations, time-resolved Raman have emerged as powerful alternatives for probing transient thermal transport dynamics. This chapter introduces the principles and applications of transient Raman characterization. Section 6.1 discusses the time-domain Raman technique, which captures the temporal evolution of temperature. Section 6.2 presents the time-domain fluorescence technique, which shares the same measurement principle but using the fluorescence signal. Section 6.3 introduces the frequency-resolved Raman, which employs modulated excitation to extract frequency-dependent thermal transport properties. Section 6.4 focuses on the energy transport state-resolved Raman approach. Together, these transient Raman techniques offer new capabilities for characterizing ultrafast heat transfer processes and understanding dynamic thermal behaviour in nanoscale systems beyond the reach of steady-state methods.

Keywords Transient Raman thermometry · Time-domain Raman technique · Frequency-resolved Raman · Non-equilibrium thermal transport

6.1 Time-domain Raman Technique

The time-domain differential Raman technique (TD-Raman) was first developed by Dr. Wang's lab at Iowa State University [1]. The development of this technique was motivated by the past transient measurement methods, which have proven to be a powerful tool for probing the thermal behaviour of materials [2, 3]. The concept of

Y. Yue, *Raman Thermometry*, https://doi.org/10.1007/978-981-95-5985-5_6

the TD-Raman is illustrated in Fig. 6.1. The method incorporates both transient and steady-state measurements. Specifically, the transient measurement consists of an excitation stage (t_e) followed by a thermal relaxation stage (t_r). Furthermore, t_r must be sufficiently long to allow the sample to fully cool before the next laser cycle begins. Several cycles are employed to obtain a sufficient Raman signal.

During the excitation period, the temperature rise and Raman response are recorded simultaneously, as illustrated in Fig. 6.1a. As the sample temperature

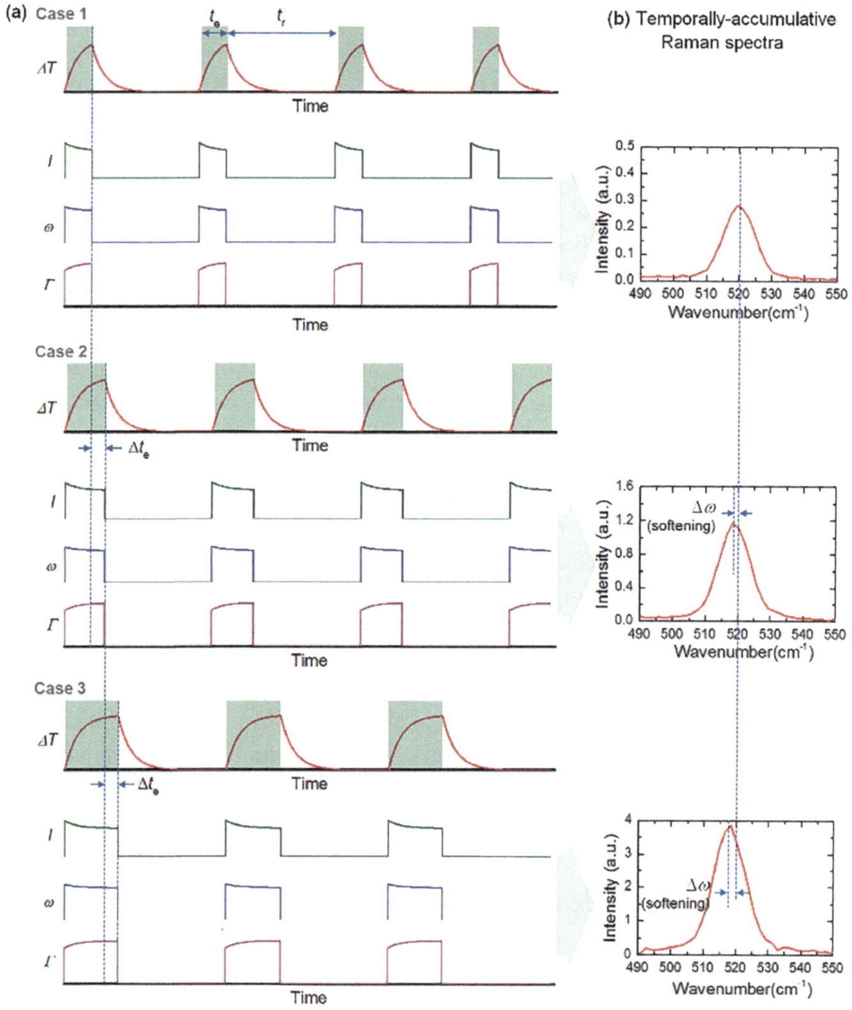

Fig. 6.1 (**a**) Timing diagrams of the laser pulse and the resulting temperature changes, accompanied by instantaneous changes in Raman peak intensity (*I*), peak shift (*ω*) and linewidth (*Γ*). (**b**) Temporally accumulated Raman spectra of one cycle for the three cases. (Reprinted from ref. [1], Copyright(2015), with permission from the Optical Society)

increases, the Raman intensity decreases, the peak shifts to a lower wavenumber and the linewidth broadens. During the thermal relaxation period (t_r), the laser is switched off and no Raman signal is recorded. Comparing the two cases, the excitation period in Case 2 is slightly longer than in Case 1, while the relaxation periods remain identical. The extended excitation in Case 2 leads to a higher temperature rise, producing a corresponding change in the Raman spectrum. Figure 6.1b presents the temporally accumulated Raman spectra for the two cases. Case 2 exhibits higher Raman intensity compared to Case 1, along with a noticeable shift in the Raman peak position. Through examination of the Raman signal variations and subsequent physical data analysis, the sample's temperature profile can be reconstructed, allowing the thermal diffusivity to be evaluated [4].

A carbon nanotube (CNT) is used to demonstrate the effectiveness of this technique. The carbon nanotube is suspended between heat sinks to enable one-dimensional thermal characterization, as shown in Fig. 6.2a. A continuous laser is modulated to generate square-wave pulses with the durations of the on and off states being variable. As shown in Fig. 6.2b, the laser beam comprises an excitation period (t_e) followed by a relaxation period (t_r). During excitation, the sample is heated and the resulting Raman scattering is recorded. Here, the Raman signal represents an integrated value obtained from contributions at different temperatures throughout

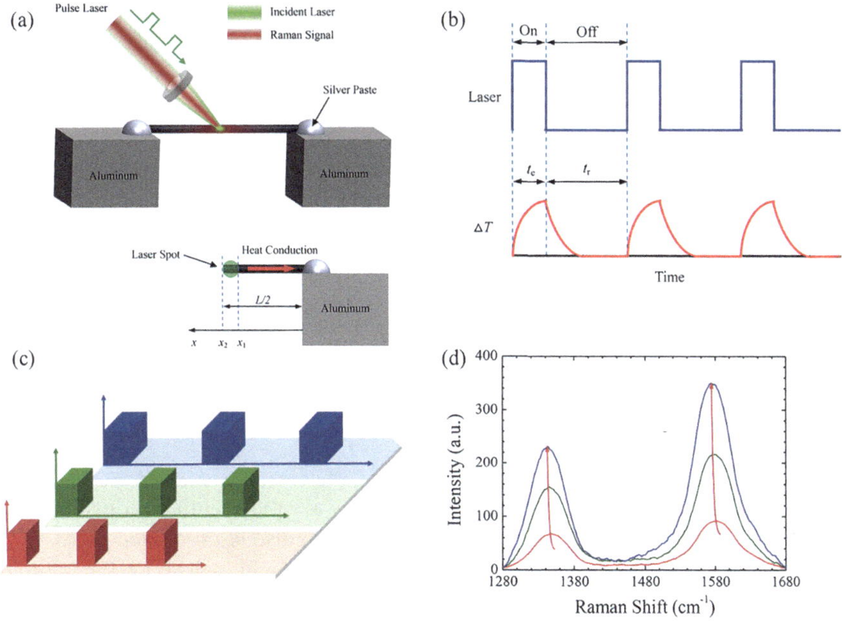

Fig. 6.2 (**a**) Schematic of the time-domain Raman technique for one-dimensional thermal characterization. (**b**) Temperature response during pulse laser heating. (**c**) Examples of laser modulation and on-off duration setting. (**d**) Raman shifts results observed under different laser pulse heating in (**c**). (Reprinted from ref. [5], Copyright(2016), with permission from Elsevier)

the laser heating period. The thermal relaxation period must be sufficiently long to allow the sample to fully return to ambient temperature before the next cycle. Multiple "heating and cooling" cycles are performed (by adjusting the integration time) to obtain the Raman spectrum. If the laser spot is larger than the sample diameter and the length-to-diameter ratio is sufficiently large, the one-dimensional thermal transport model is applicable to the laser heating experiment, as described by [1]:

$$\rho c_P \frac{\partial \theta}{\partial t} = k \frac{\partial^2 \theta}{\partial x^2} + q_0 \tag{6.1}$$

where ρ and c_P are the density and specific heat of the sample, respectively. θ is the temperature rise, k is thermal conductivity, x is the distance from the heat sink, q_0 is the heat generation per unit volume induced by the laser heating.

Considering the symmetry, the temperature rise averaged over the heated region can be expressed as follows [1]:

$$\bar{\theta}(t) = \frac{2q_0 L^3}{k(x_2 - x_1)} \sum_{m=1}^{\infty} \frac{1}{m^4 \pi^4} \left[1 - \exp\left(-\frac{m^2 \pi^2 \alpha t}{L^2}\right)\right] \left(\cos\frac{m\pi}{L} x_1 - \cos\frac{m\pi}{L} x_2\right)^2 \tag{6.2}$$

where α is the thermal diffusivity of the sample, the coordinates of x_1 and x_1 are illustrated in Fig. 6.2a, L represents the length of the sample. Figure 6.2b, c show three selected Raman spectra of the CNT fiber under different excitation times (t_e) in one "heating and cooling" cycle. As the excitation time increases, the Raman peaks (D band and G band) shift towards shorter wavelengths. This shift is attributed to the temperature rise at the heating point with increasing t_e. Raman intensity also increases due to the longer excitation time, which in turn increases the scattering collection time. From the measured Raman signal, the average temperature rise in the heated region can be obtained. The thermal diffusivity of the sample is subsequently calculated by fitting the normalized temperature increase to the heating time, as expressed by the following equation [1]:

$$\bar{\theta}(t) = \frac{\bar{\theta}(t)}{\bar{\theta}(t \to \infty)} = \frac{\sum_{m=1}^{\infty} \frac{1}{m^4 \pi^4} \left[1 - \exp\left(-\frac{m^2 \pi^2 \alpha t}{L^2}\right)\right] \left(\cos\frac{m\pi}{L} x_1 - \cos\frac{m\pi}{L} x_2\right)^2}{\sum_{m=1}^{\infty} \frac{1}{m^4 \pi^4} \left(\cos\frac{m\pi}{L} x_1 - \cos\frac{m\pi}{L} x_2\right)^2} \tag{6.3}$$

The continuous laser is modulated into a square-wave pulse with an adjustable duty cycle. The duty ratio: $t_e/(t_e + t_r)$ of the square-wave pulse laser is fixed at 5% in the measurements. The heating time per cycle is controlled within the range of 5–340 ms by varying the modulation frequency. The characteristic time for thermal transport along the sample can be estimated as $0.81l^2/\alpha$ [1, 3], where l is half length of the sample. With a referenced thermal diffusivity of 1.51×10^{-5} m^2/s [6], the characteristic thermal transport time in this fiber is calculated to be 53.6 ms. For the case of

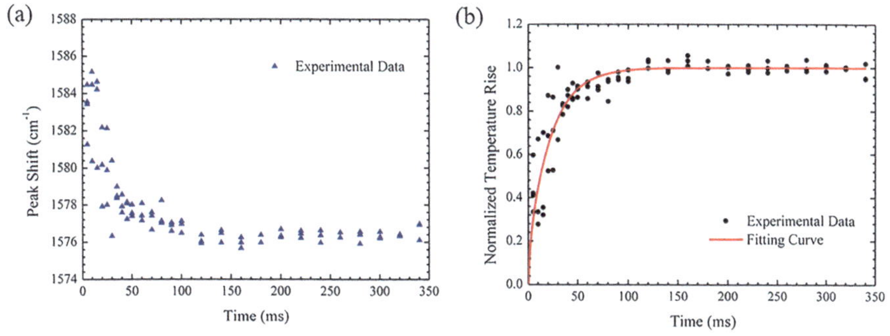

Fig. 6.3 (**a**) Experimental results of Raman shift changes as a function of laser heating duration. (**b**) Fitting results for the normalized temperature rise versus heating duration. (Reprinted from ref. [5], Copyright(2016), with permission from Elsevier)

5 ms heating experiments, the shortest relaxation time is 95 ms, ensuring that no heat accumulation occurs in the sample.

Figure 6.3a shows the measurement results when the sample is heated by a square-wave laser. During the excitation period, the average temperature is determined from the G band peak shift. The G band peak shift decreases rapidly at the onset of laser heating, indicating a fast temperature rise before thermal transfer to the electrodes. As the heating time increases, the rate of peak shift variation gradually decreases and approaches a constant value, corresponding to the sample reaching a steady-state temperature. Owing to the good linear relationship between temperature and the Raman peak shift, the normalized temperature rise of the CNT fiber can be determined from [5]:

$$\overline{\theta^*} = \frac{\omega(t) - \omega_0}{\omega_s - \omega_0} \tag{6.4}$$

where $\omega(t)$ represents the Raman shift at various heating times, ω_0 and ω_s correspond to the Raman shifts of the sample measured at room temperature and steady state, respectively. Figure 6.2b illustrates the relationship between the normalized temperature rise and heating duration. The thermal diffusivity of $1.74^{+0.20}_{-0.20} \times 10^{-5}$ m^2/s is obtained by using the least square fitting method with Eq. (6.3).

Compared with the steady-state Raman, the transient approach eliminates the need for detailed laser absorption values and additional temperature calibration experiments. Moreover, Raman measurement based on a single laser provides a simplified experimental configuration and enables rapid thermal characterization owing to its self-heating and self-probing nature. The successful measurement of a CNT fiber demonstrates the strong capability of this transient Raman and highlights its potential for broader applications.

6.2 Time-domain Fluorescence Technique

Raman thermometry is constrained to Raman active materials and mostly narrowed to some crystalline semiconductors. However, when it comes to polymers, Raman scattering is not as significant as in other crystalline structures. Especially, Raman is generally unsuitable for the thermal characterization of polymers. In addition to Raman, fluorescence is another commonly used technique that also possesses a strong temperature-dependent property. Moreover, fluorescence signal can be easily detected [7–10]. Based on the concept of time-resolved measurements, transient fluorescence spectroscopy can be used to characterize the thermal properties of polymer and other fluorescence-active materials.

A time-domain fluorescence method is developed to characterize thermophysical properties of polymers. The method is based on fluorescence thermometry of materials under periodic pulse heating. Fluorescence signal is temperature dependent in terms of photon density (signal intensity), photon frequency (wavelength), and photon decay (lifetime). Among various temperature indicators, fluorescence intensity is strong and features high sensitivity for temperature probing. Thus, based on the excited fluorescence intensity, the temperature of sample surface can be obtained under pulse laser heating. The pulse laser consists of an excitation period (t_e) followed by a relaxation period (t_r). Under laser heating, rising temperature of the sample causes a decrease of fluorescence intensity. During the off-duty period (much longer time), the sample is fully cooled down to room temperature. The fluorescence spectrum in one period of heating-cooling cycle is shown in Fig. 6.4a. As excitation times (t_e) is increased, fluorescence intensity is decreased. The temperature rise under different laser heating period can be obtained by accumulating fluorescence signal exited from all pulses. Figure 6.4b shows a significant decrease in fluorescence intensity for sample temperature from 30 °C to 70 °C.

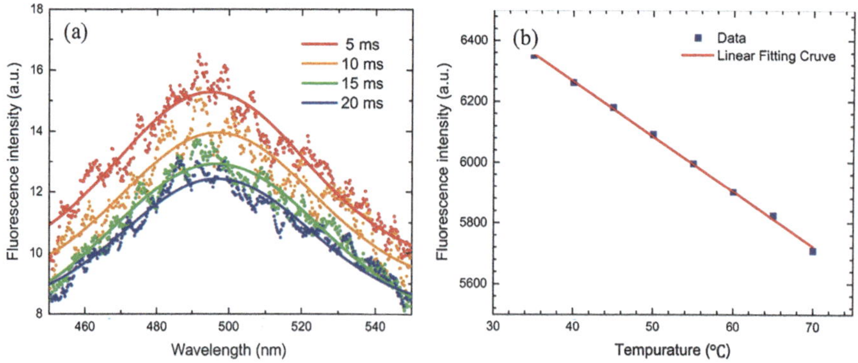

Fig. 6.4 (**a**) Typical fluorescence spectra with various heating periods. (**b**) The relationship between fluorescence intensity and temperature. (Reprinted from ref. [11], Copyright(2018), with permission from Elsevier)

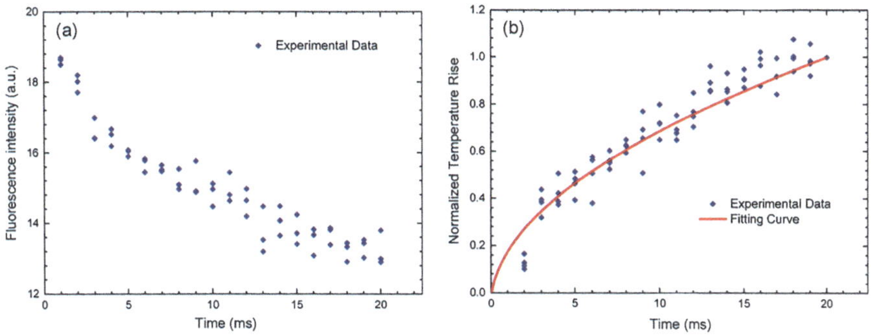

Fig. 6.5 (**a**) The relationship between fluorescence intensity and heating time. (**b**) The normalized temperature rise and the fitting curve for measuring the thermal diffusivity as 1.031×10^{-7} m^2/s. (Reprinted from ref. [11], Copyright (2018), with permission from Elsevier)

Figure 6.5a shows the relationship between fluorescence intensity and heating time. A decreasing trend for fluorescence intensity is observed over a short period time (20 ms), reflecting the temperature rise under laser heating. Based on the linear relationship between fluorescence intensity and temperature rise $I \propto T$ established during calibration, the normalized temperature rise of PVC can be expressed as: $T(t)^* = [I(t) - I_0]/[I_m - I_0]$, where I(t) is the fluorescence intensity at a given heating time, I_0 and I_{m} correspond to the fluorescence intensities at room temperature and at the maximum heating time (20 ms), respectively.

The normalized temperature rise is shown in Fig. 6.5b. Due to the low thermal conductivity and diffusivity of the polymer, a significant temperature rise occurs within a short heating period. To minimize heat accumulation, the first heating-cooling cycle is used to extract fluorescence signal in the measurement. Based on the above model, the thermal diffusivity of PVC is determined to be 1.03×10^{-7} m^2/s, which agrees well with previously reported values obtained using the laser flash method (1.09×10^{-7} m^2/s) and hot wire method (0.98×10^{-7} m^2/s) [12].

The successful measurement of PVC demonstrates that the time-domain transient fluorescence can be effectively applied to polymers. Compared with conventional techniques, this approach is simpler, requiring only a single laser, and does not necessitate prior knowledge of the laser absorption or temperature coefficient calibration either. Moreover, since most polymers exhibit strong fluorescence excitation, this method holds broad potential for application to new polymer materials.

6.3 Frequency-resolved Raman Method

The TD-Raman has an obvious technical drawback: when the heating time is extremely short, ~20 μs in the above-mentioned evaluating experiment, the integration time in the Raman spectrometer would needs to be as long as half an hour to get a sound Raman spectrum. This long integration will introduce more environmental interference to the experiment, and reduce the measurement accuracy. An effective way to solve this issue is to use the frequency-resolved Raman (FR-Raman) method [12], which is an advanced technique also developed in Wang's lab at Iowa State. The FR-Raman technique redesigns the duty of the pulsed laser based on TD-Raman technique, and extends its measuring capability to probe a very fast transient thermal transport at the scale of microseconds to nanoseconds. Meanwhile, it could provide a sound spectrum with a sufficient signal-to-noise ratio for data processing.

Figure 6.6 shows the physical principle of the FR-Raman. An amplitude-modulated square-wave laser with different frequencies (*f*) is employed to heat the sample and excite Raman signals simultaneously. Figure 6.6a shows that the durations of the laser excitation time and the thermal relaxation time are the same. The temperatures at the beginning and end of the laser excitation time are frequency-dependent. At very high frequencies, shown in Fig. 6.6b, the temperature rises in the laser excitation time and the temperature fall in the thermal relaxation time are almost negligible. This state is named as "quasi-steady state." At very low frequencies, shown in Fig. 6.6c, the laser excitation time is long enough for the temperature rising to a steady-state temperature. Furthermore, the rising period, which was much smaller than the heating period, can be neglected. As a result, the sample temperature can be seen as a constant during the excitation time. This state is termed "steady state." The temperature values at these two states are then taken as T_{qs} and T_s. Instead of absolute temperature, the temperature rise θ is a more useful way to characterize the laser heating effect. By subtracting the room temperature T_r, we can

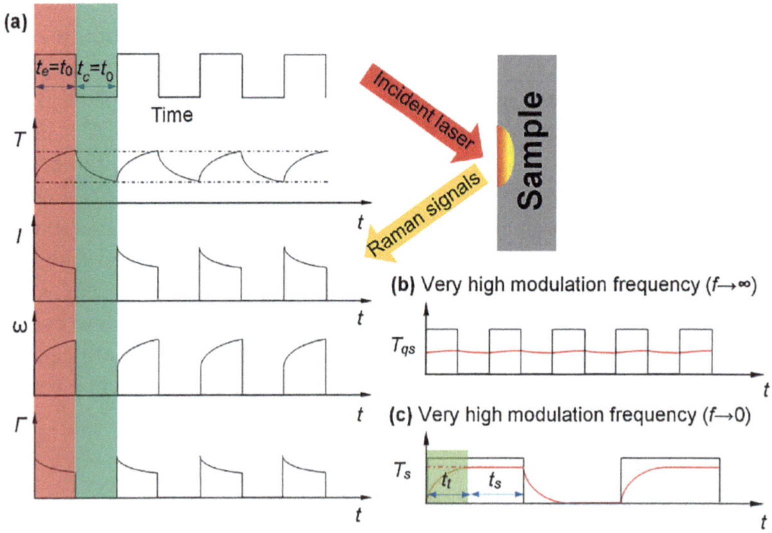

Fig. 6.6 (**a**) Concept and physics of FR-Raman. (**b**) The quasi-steady state under very high *f*. (**c**) The steady state under very low *f*. (Reprinted from ref. [13], Copyright (2016), with permission from the Optical Society)

get the temperature rise at quasi-steady state and steady state, respectively, which is $\theta_{qs} = T_{qs} - T_r$ and $\theta_s = T_s - T_r$. The relation between θ_{qs} and θ_s can be readily proved as $\theta_{qs} = \theta_s/2$. As the heating effect increases with the decrease of frequency, Raman intensity decreases and Raman peak redshifts. The Gaussian function is used to fit the Raman peaks to obtain precise Raman properties. The variation of these properties against the modulation frequency can be fitted to determine the thermal diffusivity of a sample.

Wang et al. [13] used this method to measure the thermal diffusivity of a microscale Si cantilever. As shown in Fig. 6.7a, a continuous wave laser is amplitude modulated using a function generator. The temporal profile of the laser is that of a square wave with a rise time of 130 ns. It is integrated into the original optical path of a commercial confocal Raman system. The modulated laser beam is focused by a 4× objective lens to a spot size of 31.4 × 65.3 μm. To heat the Si cantilever tip, but at same time minimize the thermal stress and avoid the possible material damage, a low laser power of 7.9 mW (before modulation) is used. An f range from 100 kHz to 17 Hz is sufficient to capture the quasi-steady-state and steady-state response. Figure 6.7b gives several examples of Si Raman peaks at different f.

The fitted variation of Raman intensity, wavenumber, and linewidth against the modulation frequency (f) is used to extract thermal diffusivity. The process for these Raman spectrum properties changing from the steady state to the quasi-steady state is determined by the thermal diffusivity of the sample. To simplify the heat conduction model, the laser heating area takes a square shape with the same size as the real case. The spatially averaged temperature rise of the heating area is developed based on the step-laser heating model [1]. The square-wave laser heating can be expressed as subsequent addition and subtraction of step-laser heating beams. The average temperature rise of the heating area is [13].

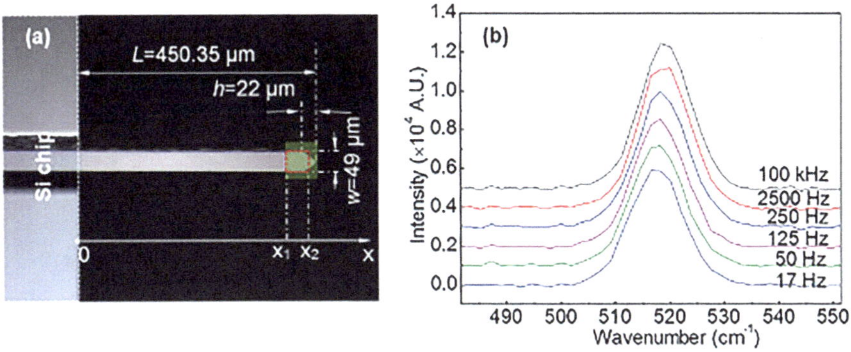

Fig. 6.7 (**a**) Optical image of the tipless Si cantilever used in the experiment. The green area is the laser spot. (**b**) Typical Si Raman peaks at selected f. (Reprinted from ref. [13], Copyright (2016), with permission from the Optical Society)

$$\bar{\theta}(t)_{\text{pulse}} = C_0 \sum_{m=1}^{\infty} C_{\text{m}} \left[1 - e^{-m^2\pi^2\alpha t/4l^2} / \left(1 + e^{-m^2\pi^2\alpha/8f_0 l^2} \right) \right] \tag{6.5}$$

where $C_m = \{[1 - (-1)^m] \cos [m\pi x_1/(2l)]\}^2/(m^4\pi^4)$ and $C_0 = 8gl^3/[(l - x_1)k]$. $l = L - h/2$ is the length of the shaped refined cantilever. $x_1 = 418.95$ μm and $x_2 = l = 438.88$ μm defines the laser heating area. g represents the laser heating source. k and α is the thermal conductivity and thermal diffusivity of the sample. f_0 is the laser modulation frequency. The maximum temperature rise $\bar{\theta}(t)_{\text{pulse\_max}}$ can be determined from this equation when $t = 1/(2f_0)$ and $1/(2f_0)$ approaches infinity ($t_0 \to \infty$) (as shown in Fig. 6.8).

Based on the above equation, the theoretical Raman spectrum at different modulation frequencies can be calculated. The evolution profile of Raman properties: intensity, Raman wavenumber, and emission, against frequency are reconstructed and used for fitting to determine the thermal diffusivity. The thermal diffusivity is determined as 9.57×10^{-5} m^2/s, 11.00×10^{-5} m^2/s and 9.02×10^{-5} m^2/s via fitting Raman intensity, wavenumber, and total Raman emission, respectively. In FR-Raman, the integration time remains constant regardless of the laser modulation speed. This helps to significantly weaken the disturbance from surroundings. This great advantage may improve the possible temporal resolution to faster than nanoseconds in Raman-based temperature response. This will provide a great potential to directly measure the interface thermal conductance.

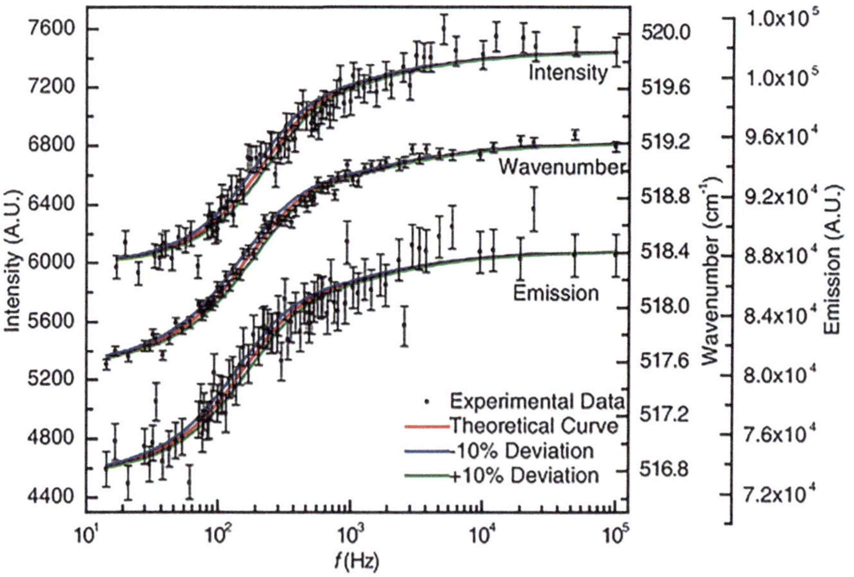

Fig. 6.8 Variation of Raman intensity, wavenumber and total Raman emission against *f*. (Reprinted from ref. [13], Copyright (2016), with permission from the Optical Society)

6.4 Energy Transport State-resolved Raman

In the laser heating processes, the hot carriers diffuse out of the irradiated area until they recombine with holes, leading to an enlarged heating area. This heating area is strongly interrelated with the hot carrier diffusion length [14], which could be comparable to the laser spot size. Dr. Wang's lab also developed the technique based on Raman to simultaneously determine the hot carrier diffusion coefficient (D) and interfacial thermal resistance (R) [14–18]. They named this technique as a picosecond (ps) energy transport state-resolved Raman (ET-Raman), which realized a ps heat transfer scenario to directly probe the diffusion of hot carriers and their contribution to energy transfer. Figure 6.9 shows the schematic of ET-Raman for measurement of D and R. The steady-state heating is accomplished by applying a CW (532 nm) laser. Under CW laser irradiation, different objective lenses (20×and 100×) were employed to differentiate the effect of D and R. By applying a picosecond laser (532 nm, the pulse duration is 13 ps, the repetition rate is 48.2 MHz), a near zero-transport state is constructed. Within such a short laser pulse (13 ps), no electron-hole recombination occurs. The temperature rise under ps laser heating is more determined by volumetric heat capacity (ρC_p) rather than the R and D.

For steady state, the generation and diffusion of heat and hot carriers are governed by two partial differential equations ($\partial N/\partial t = 0$) [19, 20]. The first one is the carrier diffusion equation to determine the hot carrier concentration $\Delta N(r, t)$ cm^{-3} [14]:

$$D\nabla^2 \Delta N - \frac{\Delta N}{\tau} + \frac{\partial n_0}{\partial T_{\mathrm{CW}}} \frac{\Delta T_{\mathrm{CW}}}{\tau} + \Phi a = 0 \tag{6.6}$$

where D (cm^2/s), τ (s), and Φ (photons/cm^{3}s) are the carrier diffusion coefficient, the electron–hole recombination time of the sample and the incident photon flux of the laser source. a (cm^{-1}) is the optical absorption coefficient of the sample. n_0 is the

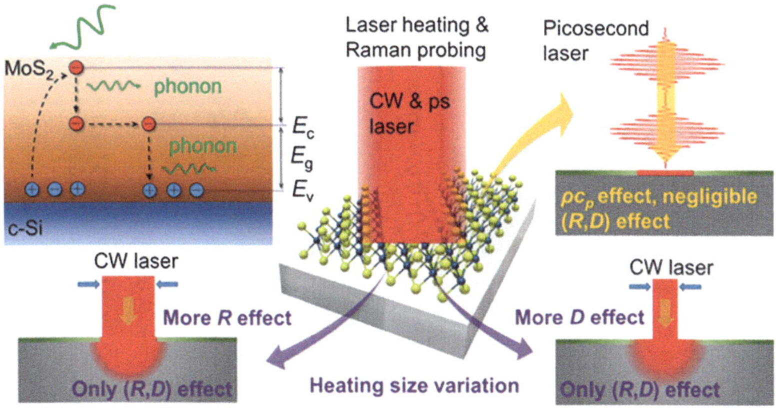

Fig. 6.9 Artistic representation of the ET-Raman experiment concept. (Reprinted from ref. [14], Copyright (2017), with permission from American Chemical Society)

equilibrium free-carrier density at temperature T. The second equation is the thermal diffusion equation, which involves the free carrier density [14]:

$$k\nabla^2 \Delta T_{CW} + \left(h\nu - E_g\right)\Phi a + \frac{E_g \Delta N}{\tau} = 0 \tag{6.7}$$

where $\Delta T_{CW}(r,t)$ (K), k (W/mK) are the temperature rise in the steady state heating, the thermal conductivity of sample, and $h\nu$ is the photon energy of the laser source. Due to the hot carrier diffusion effect, the real heating area will be larger than the laser irradiating area, and is highly related to the hot carrier diffusion length ($L_D = \sqrt{\tau D}$). As a result, when the laser spot size is large enough, the hot carrier diffusion will have a negligible effect on the heating area [21].

To rule out the effect of laser absorption coefficient and Raman temperature coefficient, a normalized Raman shift power coefficient (RSC) is defined as $\Theta_1 = \chi_{CW1}/\chi_{ps}$ and $\Theta_2 = \chi_{CW2}/\chi_{ps}$. Although χ_{CW1}, χ_{CW2}, and χ_{ps} are all influenced by the Raman temperature coefficient and the laser absorption in the sample, this normalized RSC completely rules out the effect of laser absorption and Raman temperature coefficients. Θ_1 and Θ_2 are only a function of the 2D and substrate materials' ρC_p, R, and D. By solving the 3D heat conduction and hot carrier diffusion equations numerically, the D and R of a 2D material can be determined (Fig. 6.10).

One critical feature of the ET-Raman is that the normalized RSC eliminates the need for temperature rise and laser absorption in interface thermal conductance determination. Another advantage is that it significantly magnifies the effect of G of 2D materials on a substrate of low thermal conductivity. This ET-Raman technique could also be applicable to other sample structures, such as suspended 2D materials. For this structure, the absorbed laser energy could only dissipate along the in-plane direction. Additionally, the sample thermal relaxation time will be longer and there could be a heat accumulation effect by laser pulses. Therefore, the sample could be

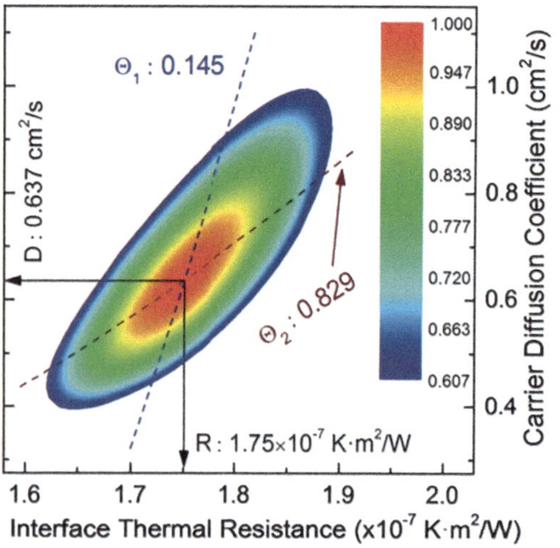

Fig. 6.10 3D numerical modelling results for determining D and R. (Reprinted from ref. [14], Copyright (2017), with permission from American Chemical Society)

easily destroyed during the first several laser pulses. However, by controlling the laser to have a longer cooling time between pulses, ET-Raman can still be applied to characterize suspended 2D materials.

6.5 Summary

This chapter presents transient Raman methods for characterizing the thermal properties of materials. By using two different laser heating sizes in steady-state laser heating, along with a pulsed laser (nanosecond or picosecond) heating state, the in-plane thermal conductivity and hot carrier diffusion coefficients can be determined. Furthermore, this approach eliminates the need for measuring absorbed laser heating power and absolute temperature rise, which could introduce significant uncertainties in previous measurements. As a result, it provides a much more accurate understanding of interface energy coupling and hot carrier diffusion.

References

1. Xu, S., Wang, T.Y., Hurley, D., Yue, Y.N., Wang, X.W.: Development of time-domain differential Raman for transient thermal probing of materials. Opt. Express. **23**(8), 10040–10056 (2015)
2. Xi, S.S., Zhang, N., Huang, X.N., Huang, D.Z., Zhang, Z.F., Guo, Y.Z., Yue, Y.N.: Upper and lower limits of thermal diffusivity for Ti3C2Tx MXene fiber. Int. J. Thermophys. **46**(9), 134 (2025)
3. Guo, J.Q., Wang, X.W., Wang, T.: Thermal characterization of microscale conductive and nonconductive wires using transient electrothermal technique. J. Appl. Phys. **101**(6), 063537 (2007)
4. Li, Q.Y., Zhang, X., Takahashi, K.: Variable-spot-size laser-flash Raman method to measure in-plane and interfacial thermal properties of 2D van der Waals heterostructures. Int. J. Heat Mass Transf. **125**, 1230–1239 (2018)
5. Li, C., Xu, S., Yue, Y., Yang, B., Wang, X.J.: Thermal characterization of carbon nanotube fiber by time-domain differential Raman. Carbon. **103**, 101–108 (2016)
6. Li, M., Li, C.Z., Wang, J.M., Xiao, X.H., Yue, Y.A.: Parallel measurement of conductive and convective thermal transport of micro/nanowires based on Raman mapping. Appl. Phys. Lett. **106**(25), 253108 (2015)
7. Bur, A.J., Vangel, M.G., Roth, S.: Temperature dependence of fluorescent probes for applications to polymer materials processing. Appl. Spectrosc. **56**(2), 174–181 (2002)
8. Gota, C., Uchiyama, S., Yoshihara, T., Tobita, S., Ohwada, T.: Temperature-dependent fluorescence lifetime of a fluorescent polymeric thermometer, poly(-isopropylacrylamide), labeled by polarity and hydrogen bonding sensitive 4-sulfamoyl-7-aminobenzofurazan. J. Phys. Chem. B. **112**(10), 2829–2836 (2008)
9. Kujawa, P., Tanaka, F., Winnik, F.M.: Temperature-dependent properties of telechelic hydrophobically modified poly(-isopropylacrylamides) in water: evidence from light scattering and fluorescence spectroscopy for the formation of stable mesoglobules at elevated temperatures. Macromolecules. **39**(8), 3048–3055 (2006)
10. Ye, F.M., Wu, C.F., Jin, Y.H., Chan, Y.H., Zhang, X.J., Chiu, D.T.: Ratiometric temperature sensing with semiconducting polymer dots. J. Am. Chem. Soc. **133**(21), 8146–8149 (2011)
11. Wu, H., Cai, K., Zeng, H., Zhao, W., Xie, D., Yue, Y., Xiong, Y., Zhang, X.J.: Time-domain transient fluorescence spectroscopy for thermal characterization of polymers. Appl. Therm. Eng. **138**, 403–408 (2018)

12. Wang, T.Y., Xu, S., Hurley, D.H., Yue, Y.N., Wang, X.W.: Frequency-resolved Raman for transient thermal probing and thermal diffusivity measurement. Opt. Lett. **41**(1), 80–83 (2016)
13. Wang, T., Xu, S., Hurley, D.H., Yue, Y., Wang, X.J.: Frequency-resolved Raman for transient thermal probing and thermal diffusivity measurement. Opt. Lett. **41**(1), 80–83 (2015)
14. Yuan, P.Y., Wang, R.D., Tan, H., Wang, T.Y., Wang, X.W.: Energy transport state resolved raman for probing interface energy transport and hot carrier diffusion in few-layered MoS2. ACS Photon. **4**(12), 3115–3129 (2017)
15. Yuan, P.Y., Wang, R.D., Wang, T.Y., Wang, X.W., Xie, Y.S.: Nonmonotonic thickness-dependence of in-plane thermal conductivity of few-layered MoS2: 2.4 to 37.8 nm. Phys. Chem. Chem. Phys. **20**(40), 25752–25761 (2018)
16. Yuan, P.Y., Liu, J., Wang, R.D., Wang, X.W.: The hot carrier diffusion coefficient of sub-10 nm virgin MoS2: uncovered by non-contact optical probing. Nanoscale. **9**(20), 6808–6820 (2017)
17. Yuan, P.Y., Tan, H., Wang, R.D., Wang, T.Y., Wang, X.W.: Very fast hot carrier diffusion in unconstrained MoS2 on a glass substrate: discovered by picosecond ET-Raman. RSC Adv. **8**(23), 12767–12778 (2018)
18. Hunter, N., Azam, N., Zobeiri, H., Wang, R.D., Mahjouri-Samani, M., Wang, X.W.: Interfacial thermal conductance between monolayer WSe2 and SiO2 under consideration of radiative electron-hole recombination. Acs Appl. Mater. Inter. **12**(45), 51069–51081 (2020)
19. Nestoros, M., Forget, B.C., Christofides, C., Seas, A.: Photothermal reflection versus temperature-quantitative-analysis. Phys. Rev. B. **51**(20), 14115–14123 (1995)
20. Christofides, C., Othonos, A., Loizidou, E.: Influence of temperature and modulation frequency on the thermal activation coupling term in laser photothermal theory. J. Appl. Phys. **92**(3), 1280–1285 (2002)
21. Radisavljevic, B., Radenovic, A., Brivio, J., Giacometti, V., Kis, A.: Single-layer MoS2 transistors. Nat. Nanotechnol. **6**(3), 147–150 (2011)

Chapter 7
Tip-enhanced Raman for Nanoscale Thermal Probing

Abstract Accurate temperature measurement at nanometer scales is critical for understanding localized energy dissipation, yet remains difficult due to limited spatial resolution and the complexity of heat transport at small dimensions. Section 7.1 outlines the fundamental challenges in nanoscale temperature measurement. Section 7.2 provides an overview of existing nanoscale thermal probing techniques, including scanning thermal microscopy. Section 7.3 introduces the principles and instrumentation of tip-enhanced Raman spectroscopy. Section 7.4 focuses on the development of tip-enhanced Raman thermometry (TERT), which combines the spatial precision of TERS with the temperature sensitivity of Raman thermometry to achieve localized thermal measurements with sub-10 nm spatial resolution. Finally, Sect. 7.5 discusses ballistic thermal transport probed by TERT, providing new insights into non-diffusive phonon dynamics in semiconductors.

Keywords Tip-enhanced Raman thermometry (TERT) · Nanoscale thermal probing · Ballistic heat transport · Near-field spectroscopy

7.1 Challenges in Nanoscale Temperature Measurement

Performing temperature measurements at nanoscale dimensions poses significant challenges, mainly because complex physical phenomena emerge at the nanoscale, such as the size effect and ballistic thermal transport. These factors complicate the design of nanoscale probes and at the same time reduce the accuracy of temperature readings. As device dimensions continue to shrink, heat transfer behavior gradually deviates from macroscopic models. Conducting thermal characterization at this scale reveals new physical phenomena while providing valuable insights with important engineering applications. From an application perspective, nanoscale temperature measurement shows broad prospects across multiple fields. In the semiconductor industry, the linewidth of integrated circuits has already fallen below 50 nm, and the reliability of devices under high power density increasingly depends

Y. Yue, *Raman Thermometry*, https://doi.org/10.1007/978-981-95-5985-5_7

on precise monitoring of local temperature [1]. In electronic and optoelectronic devices, nanoscale temperature measurement is essential for monitoring localized heating effects that directly impact device reliability and performance. For nanoscale device components such as nanowires, thin films, and heterostructures, localized thermal probing reveals critical information about heat dissipation pathways, which directly influences device efficiency and stability.

Nanoscale temperature measurement is crucial for studying local thermal responses under laser heating, especially in near-field effects with feature sizes below 100 nm [2, 3]. These measurements improve understanding of micro- and nanoscale heat transfer mechanisms and support the development of advanced manufacturing technologies [4]. Nanoscale temperature measurement enables the detection of highly localized hotspots and thermal gradients, which are averaged out in macroscopic measurements. Such information is essential for analyzing heat dissipation in nanostructured materials. In these materials, interfaces, defects, and surface roughness often play a dominant role in thermal transport. Additionally, understanding temperature distributions at the nanoscale helps optimize material composition and device architecture for improved thermal management. It also helps calibrate and validate computational models for nanoscale heat transport. This process bridges experimental observations with theoretical analysis. Overall, incorporating these detailed insights strengthens the interpretation of nanoscale thermal phenomena and enhances the reliability of measurements across diverse application scenarios.

7.2 Overview of Nanoscale Thermal Probing Techniques

With the rapid advancement of nanotechnology and the continued miniaturization of semiconductor devices, a wide range of nanoscale systems has emerged [5]. Understanding their thermal behavior is crucial, as efficient heat management becomes increasingly difficult when device dimensions approach the mean free path of energy carriers [6]. At the nanoscale, conventional heat transfer mechanisms become less effective, leading to localized heat accumulation that can degrade device performance and shorten operational lifetime. Accurate measurement of local temperatures at the nanoscale is therefore essential for characterizing thermal performance and guiding device design. Various experimental approaches have been developed to probe nanoscale temperatures, including scanning probe–based methods, fluorescence-based techniques, and Raman thermometry, each offering distinct advantages for studying localized thermal phenomena.

The design of nanoscale temperature probes involves accurately locating nanoscale targets and precisely measuring their temperature. The development of scanning tunneling microscopy (STM) made atomic-scale surface imaging possible and opened opportunities for temperature mapping. However, STM-based measurements require strict control of the probe–sample gap and can only be applied to metallic samples, which limits their scope. Scanning thermal microscopy (SThM), later developed on the basis of atomic force microscopy (AFM), overcomes these limitations [7]. By mounting a temperature sensor at the tip of the cantilever and maintaining a stable gap through atomic force, SThM enables direct probing of surface temperature at the nanoscale. In SThM, heat transfer between the tip and the sample is complex, involving mainly solid–solid conduction and air conduction, while radiation is usually negligible. To minimize uncertainties arising from surface roughness, a liquid film is commonly applied at the probe tip to enhance thermal contact. Over the past two decades, SThM has emerged as a powerful tool for nanoscale temperature mapping and quantitative characterization of material thermal properties. Its spatial resolution is limited by the fabrication of probes and sensors, but advances in nanofabrication and lithography have improved the resolution from the 100 nm scale to about 10 nm. At the same time, the technique shows high accuracy and temporal performance, with temperature uncertainty as low as 10^{-3} K and time resolution below 1 ms. However, its performance is still restricted by the response speed and stability of thermal probes.

Fluorescence thermometry is a non-contact and non-destructive technique that has been widely applied to nanoscale thermal imaging in recent years [8]. Its principle relies on the temperature dependence of fluorescence signals: with increasing temperature, fluorescence intensity decreases, the emission peak undergoes a redshift, and the peak width broadens. By analyzing the linear relationships between temperature and parameters such as emission wavelength, intensity, and peak width, precise temperature calibration can be achieved [9]. Quantum dots are commonly employed as fluorescent probes, typically with sizes ranging from a few nanometers to over 10 nm, enabling high spatial resolution temperature measurements at the nanoscale [10]. In addition to quantum dots, inorganic phosphors, organic dyes, and other nanoparticles can also serve as temperature-sensing materials, with preference given to systems exhibiting stable fluorescence intensity and minimal environmental interference. Advances in this field have allowed fluorescence thermometry to overcome the optical diffraction limit, achieving spatial resolutions below 100 nm and even as low as 10–20 nm, together with temperature resolutions better than 0.01 K.

As an important optical technique, micro-Raman thermometry is widely applied in the characterization of semiconductor materials and the investigation of nanoscale thermal phenomena [11]. The spatial resolution of micro-Raman thermometry is typically limited by the laser focal spot size. Resolving temperature variations at the nanoscale necessitates the use of nanostructures as localized thermal probes. Various nanomaterials [12], such as single-walled carbon nanotubes, silicon nanowires, and graphene, serve as effective platforms for Raman thermometry due to the high temperature sensitivity of their characteristic Raman peaks. In particular, the atomic thickness and exceptional optical response of graphene enable temperature measurements with near sub-nanometer resolution along the thickness direction. Raman thermometry based on nanomaterials overcomes the diffraction limit of conventional optics, providing a new approach for high-precision thermal characterization. It holds significant potential for probing interfacial thermal resistance, localized heat transport, and intrinsic thermal properties of materials, and is regarded as a key approach for advancing temperature measurement at the nanoscale.

7.3 Tip-enhanced Raman Spectroscopy

Tip-enhanced Raman Spectroscopy (TERS) combines Raman spectroscopy with scanning probe microscopy (SPM) to achieve a spatial resolution beyond the diffraction limit of conventional Raman spectroscopy. TERS has become an important tool for optical and thermal characterization of electronic devices and functional materials at the nanoscale (Fig. 7.1). In TERS, a sharp metallic tip (typically gold- or silver-coated) is precisely positioned near the sample surface. When the tip is illuminated by a laser, localized surface plasmons are excited, generating a highly confined electromagnetic field at the nanoscale region of the tip apex. This localized field enhancement significantly amplifies the Raman signal of molecules directly beneath the tip, enabling spatial resolution at the nanoscale. The signal enhancement in TERS primarily arises from two mechanisms. The first is electromagnetic enhancement, generated by the localized surface plasmon resonance of the metallic tip, which strongly confines the electromagnetic field at the apex. The second is chemical enhancement, involving charge-transfer interactions between the tip and the sample. These mechanisms act synergistically, allowing Raman signals to be amplified by several orders of magnitude under ideal conditions, even enabling single-molecule detection.

Compared with conventional Raman spectroscopy, TERS offers several distinct advantages. First, TERS can provide chemical information at the nanoscale with spatial resolution beyond the diffraction limit, which is essential for studying heterogeneous catalysts, polymer blends, two-dimensional materials, and semiconductor nanostructures. Second, TERS facilitates real-time measurements under diverse environments, such as air, liquid, and vacuum, providing a powerful means to study dynamic nanoscale processes. Third, by controlling the incident light polarization and the tip–sample distance, TERS can selectively enhance specific vibrational modes, enabling detailed analysis of molecular orientation and local chemical

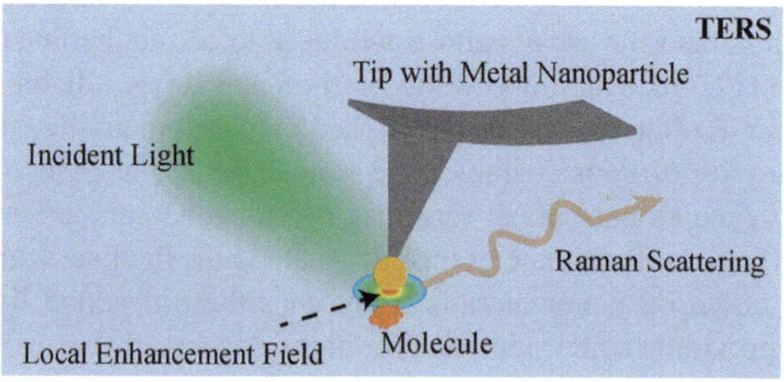

Fig. 7.1 Schematic illustration of TERS

environments. TERS has found applications across materials science, catalysis, nanoscale thermal studies, and biological systems. In materials science, it can be used to characterize local defects, grain boundaries, and stress distributions in two-dimensional materials such as graphene and transition metal dichalcogenides, which critically affect electronic and optical properties. In the field of catalysis, TERS enables the direct observation of reaction intermediates on nanoparticle surfaces, providing experimental insight into reaction mechanisms and guiding catalyst design and optimization. In biological systems, TERS allows nanoscale characterization of single molecules or protein conformations, revealing molecular interactions that are often hidden in ensemble measurements. Furthermore, TERS demonstrates unique capabilities in nanoscale thermal and photochemical studies, enabling spatially resolved analysis of local heating and energy transfer processes, and allowing direct observation of nanoscale thermodynamic phenomena.

7.4 Development of Tip-enhanced Raman Thermometry

Building on the principles and capabilities of TERS, the author combined the Raman thermometry concept and develop the new technique called Tip-enhanced Raman Thermometry (TERT), which enables nanoscale thermal measurements. While TERS primarily relies on plasmonic enhancement at the tip apex to amplify Raman signals for chemical analysis, TERT builds on this principle by using the tip to generate nanoscale hotspots on the sample surface and temperature measurement based on the Raman-temperature correlation. Combined with the tip-enhanced electromagnetic field, this approach allows the study of heat transfer at the nanoscale. This section addresses the experimental setup and inherent challenges of sub-10 nm thermal measurements.

Accurate positioning of the laser and precise adjustment of its polarization are essential for temperature quantification. Moreover, the observed temperature variations under different laser configurations underscore the critical influence of the near-field effect in localized heating at the tip apex. Figure 7.2 illustrates the overall setup for near-field heating and sub-10 nm temperature detection. In this configuration, the Raman spectrometer is positioned facing the AFM, with the laser focused on the tip.

The silicon substrate is beneath the Au-coated silicon tip, where the near-field enhancement generates localized heating. Temperature measurement via Raman thermometry relies on a well-established correlation between the Raman spectral features and the substrate temperature. The system is calibrated in advance using the silicon substrate before conducting sub-10 nm thermal experiments. To ensure that the measured temperature originates from the silicon substrate under near-field optical heating rather than the AFM tip, the silicon tip is coated with a 20 nm thick gold (Au) layer. Given that the optical absorption depth of Au at the Raman excitation wavelength is around 10 nm, the contribution of Raman signal from the tip is negligible.

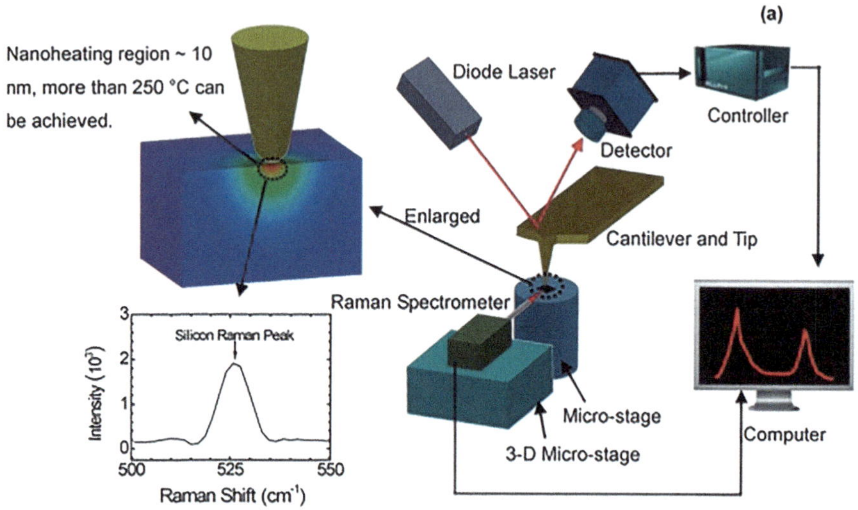

Fig. 7.2 Schematic of the experimental setup for tip-enhanced Raman thermal measurement. (Reprinted from ref. [3], Copyright (2011), with permission from American Chemical Society)

During the experiment, the laser polarization is varied at three angles (15°, 30°, and 75° relative to the tip axis) to generate distinct near-field heating conditions. The incident laser is scanned vertically using a 3-D micro stage to selectively irradiate different regions of the tip, producing distinguishable heating effects. Due to the small projected area of the tip along the laser propagation direction, the scattered intensity remains low when near-field heating is effectively induced. Conversely, high scattered intensity indicates that the laser predominantly illuminates the larger upper portion of the cantilever rather than the tip apex. To maintain consistency with the calibration procedure, all Raman spectra of the substrate were fitted using Gaussian functions. Figure 7.3 presents representative spectra for the three laser polarization angles along with their corresponding fits, demonstrating excellent agreement between the measured data and the fitted curves. Substrate temperatures were determined by correlating each experimental Raman spectrum with the calibration results.

The variation in substrate temperature under different laser–tip configurations, as inferred from the scattered laser intensity, for the three polarization angles is shown in Fig. 7.3. Prior to performing near-field Raman measurements, attempts were made to detect the surface Raman signal with the AFM tip retracted from the substrate, but no signal was observed. Accurate characterization of near-field heating and Raman response requires maintaining the tip–substrate gap within a few nanometers, to capture the high sensitivity of the system. The measurements revealed that, in the sub-10 nm regime, the silicon temperature reached up to 250 °C for laser polarizations of 15° and 30° relative to the tip axis. A sharp temperature jump was observed at the scattered light from the AFM tip. This confirms that the elevated

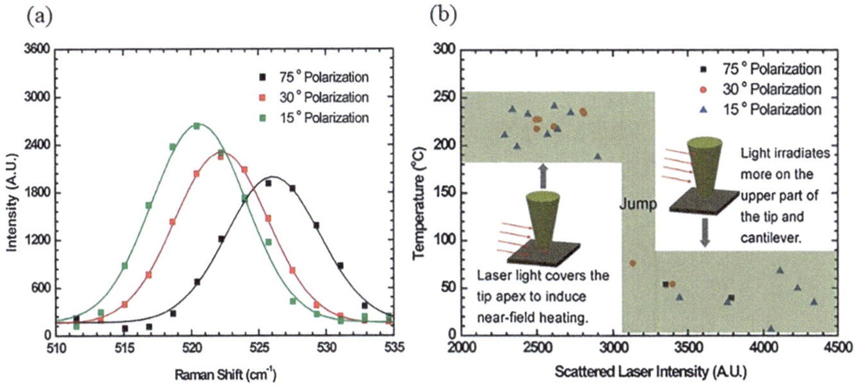

Fig. 7.3 Raman spectra and the corresponding substrate temperatures as a function of scattered laser intensity. (**a**) Typical Raman peaks at three different polarization angles; (**b**) temperature rise as laser leaves the tip apex (Reprinted from ref. [3], Copyright (2011), with permission from American Chemical Society)

temperature originated from laser focusing at the tip apex. This section systematically elaborates on the advantages of TERT in nanoscale thermal measurements and its experimental implementation. By exploiting the localized strong field generated by tip enhancement, Raman signals are greatly amplified while nanoscale hotspots are simultaneously formed, enabling temperature probing with sub-10 nm resolution. The experimental results show a pronounced influence of laser polarization direction and focal position on local heating and temperature measurement accuracy. This finding confirms the critical role of near-field effects in nanoscale thermal response studies.

7.5 Ballistic Thermal Transport Probed by TERT Technique

Based on the TERT configuration, this section explores the application for GaN substrates, emphasizing the quantitative correlation between Raman shifts and local temperature and highlighting the method's potential for nanoscale thermal characterization. In these experiments, a 532 nm laser served both as a probe and a localized heating source, with the beam focused on the gold-coated AFM nanotip [13]. The highly confined laser produces a localized heating region directly beneath the tip apex, as shown in Fig. 7.4. Initial characterization of the GaN substrate was performed using Raman spectroscopy. The incident laser was directed along the normal to the film surface, revealing E_1(TO) and A_1(LO) as the prominent active phonon modes. Temperature changes were quantified based on shifts in the E_1(TO) phonon mode due to its strong signal intensity. Local temperature increases were determined with a micro-Raman spectrometer. The linear relationship between Raman peak frequency and temperature is established. For precise thermal

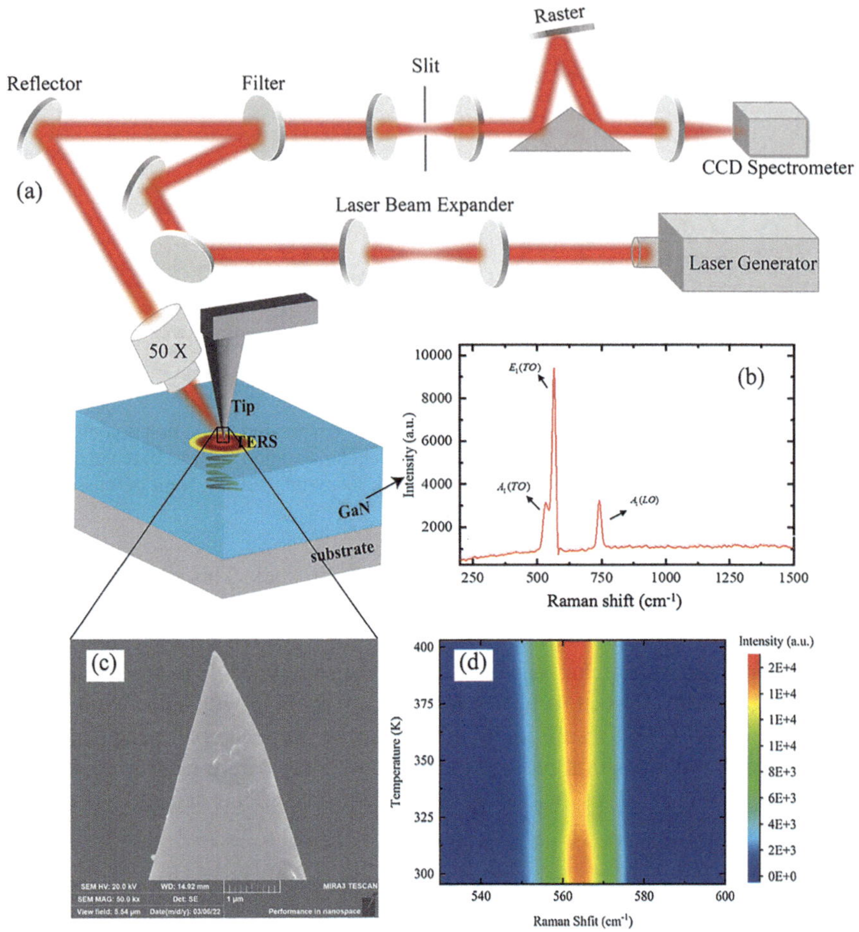

Fig. 7.4 Schematic diagram of optical alignment for thermal sensing (**a**) optical alignment for tip-Raman excitation; (**b**) Typical Raman spectrum of GaN; (**c**) SEM of nanotip; (**d**) calibration result of Raman shift [14]

measurement, the sample was mounted on a substrate holder attached to a three-axis translation stage, allowing controlled positioning of the nanotip within the target area. At the focal plane along the z-axis, the laser spot had a diameter of approximately 60 μm. The nanotip's lateral position was finely adjusted to ensure optimal laser focusing on the tip apex, thereby generating localized heating on the GaN substrate. The laser-irradiated region on the tip was precisely defined by adjusting its position, with Rayleigh scattering intensity indicating proper sample illumination. Initially, the tip was translated along the cantilever direction while maintaining fixed positions along the remaining axes. The laser was considered focused on the tip center when the Rayleigh scattering intensity reached its maximum. Subsequently,

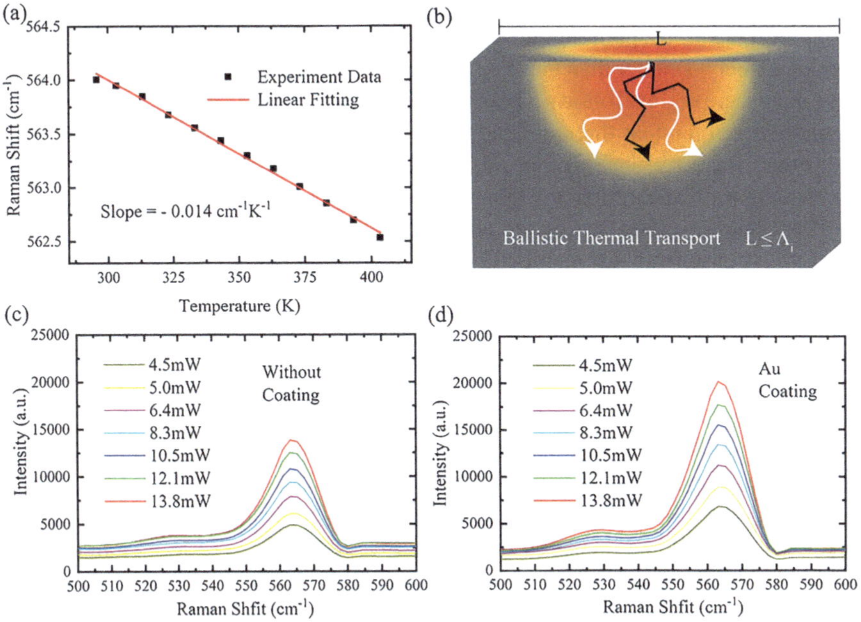

Fig. 7.5 Measurement result of tip-enhanced Raman thermometry. (**a**) Linear relationship between Raman peak position and temperature; (**b**) illustration of ballistic thermal transport; (**c**) selected GaN Raman spectra under different tips and laser powers, and (**d**) results for tip-enhanced thermal transport with gold coating [14]

the tip was adjusted along the y-axis to the location where a detectable Raman signal was obtained. A linear correlation is observed between the Raman shift of the E_1(TO) mode and the substrate temperature, as illustrated in Fig. 7.5a.

This linear trend arises from lattice expansion and contraction induced by thermal effects, which alter the equilibrium positions of atoms and consequently shift phonon energies. Balkanski et al. [15]. proposed a theoretical model describing the temperature dependence of Raman frequencies, indicating that for temperatures below 600 K, the phonon anharmonicity results in a linear relationship between frequency and temperature. The temperature coefficient of GaN was calculated as -0.014 cm^{-1}/mW during the calibration experiment, consistent with previously reported values [16–18]. The obtained temperature coefficient can subsequently be used to measure local temperature in the GaN substrate under a highly focused hotspot introduced by the gold-coated tip upon laser irradiation.

The temperature rise at the localized GaN heating spot is strongly influenced by the material's thermal conductivity. The ~10 nm tip apex generates a nanoscale hotspot via surface plasmon resonance, while at the same time amplifying the Raman signal as shown in Fig. 7.5b. The observed thermal behavior reflects ballistic heat transport. The laser-induced heated region in the GaN substrate is comparable in size to the tip apex and much smaller than the typical phonon means free path

(MFP) in GaN. In bulk GaN at room temperature, heat is primarily carried by phonons with mean free paths in the range of 100–1000 nm. The size of the laser-heated region is much smaller than these mean free paths, therefore, the energy transport in this nanoscale regime follows a ballistic mechanism consistent with Landauer's formulation for quantum thermal conduction [19, 20]. Upon laser illumination of the substrate, localized heating occurs in silicon tip systems, with the gold coating enhancing light absorption and thereby promoting ballistic thermal transport.

Representative Raman spectra were recorded using the bare silicon tip at laser powers ranging from 4.5 mW to 13.8 mW, as shown in Fig. 7.5c. With increasing laser power, the Raman-active modes exhibit softening due to enhanced local heating of the GaN substrate, which increases phonon–phonon scattering. Importantly, excitation up to 13.8 mW does not damage the sample, and the linear correlation between the GaN Raman shift and temperature is preserved. Reducing the laser power to limit heating inevitably diminishes the Raman signal-to-noise ratio, thereby complicating quantitative measurements. For the gold-coated silicon tip, higher laser powers similarly induce a red shift and stronger Raman intensity, as shown in Fig. 7.5d. The gold coating provides plasmonic enhancement. It amplifies the Raman signal and improves spatial thermal resolution. The Raman intensity of GaN beneath the coated tip is about 1.5 times higher than that under the bare tip. This observation additionally confirms the decreasing trend of the GaN Raman shift with increasing laser power, as illustrated in Fig. 7.6a.

The fitted slopes of the Raman shift–power relationship are −0.0445 and −0.0309 cm^{-1}/mW for the bare and gold-coated silicon tips, respectively. This linear decrease with laser power indicates that, under the same excitation power, thermal energy can be absorbed more efficiently, leading to a higher local temperature. This trend can be extracted from the previously established temperature–Raman shift calibration. The temperature of the GaN hotspot increases with laser power, reflecting enhanced energy absorption induced by surface plasmon resonance, as shown in Fig. 7.6b. The temperature rise caused by the gold-coated tip (~40 K) is substantially higher than that generated by the bare silicon tip (~20 K). These results

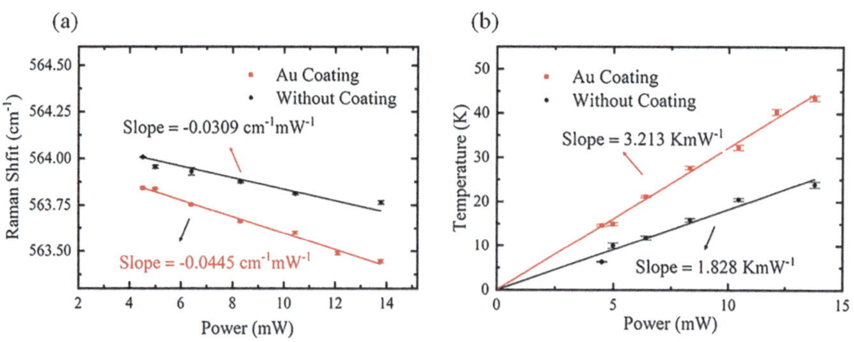

Fig. 7.6 Measurement result of tip-enhanced Raman thermometry on GaN. (**a**) Raman shift as a function of laser power; (**b**) derived temperature as a function of laser power [14]

demonstrate that TERT enables direct observation of quasi-ballistic thermal transport in GaN.

The localized electromagnetic field from tip enhancement amplifies the Raman signal and produces sub-10 nm hotspots at the tip apex, facilitating high spatial-resolution temperature measurements. Experimental results indicate that the gold-coated tip markedly improves local light absorption. Surface plasmon resonance enhances these effects and makes quasi-ballistic thermal transport more clearly observable. Overall, this study demonstrates the potential of TERT for nanoscale thermal characterization and provides an effective means to directly probe quasi-ballistic heat flow, offering a valuable experimental tool for local thermal management and the investigation of nanoscale thermal phenomena in semiconductor materials.

7.6 Summary

This chapter highlights the application of TERT for nanoscale thermal probing. Benefiting from localized electromagnetic field enhancement, TERT enables the detection of local thermal responses with sub-10 nm resolution. Experimental findings indicate that laser polarization and focal positioning substantially impact local heating and measurement accuracy. Investigations on GaN substrates further reveal quasi-ballistic heat transport, with gold-coated tips enhancing signal intensity via surface plasmon resonance. Overall, TERT provides an effective tool for probing localized thermal phenomena at the nanoscale and shows great potential for advancing the understanding of heat transport mechanisms.

References

1. Aigouy, L., Tessier, G., Mortier, M., Charlot, B.: Scanning thermal imaging of microelectronic circuits with a fluorescent nanoprobe. Appl. Phys. Lett. **87**(18), 184105 (2005)
2. Korte, F., Nolte, S., Chichkov, B.N., Bauer, T., Kamlage, G., Wagner, T., et al.: Far-field and near-field material processing with femtosecond laser pulses. Appl. Phys. A Mater. Sci. Process. **69**, S7–S11 (1999)
3. Yue, Y., Chen, X., Wang, X.: Noncontact sub-10 nm temperature measurement in near-field laser heating. ACS Nano. **5**(6), 4466–4475 (2011)
4. Malshe, A.P., Rajurkar, K.P., Virwani, K.R., Taylor, C.R., Bourell, D.L., Levy, G., et al.: Tip-based nanomanufacturing by electrical, chemical, mechanical and thermal processes. Cirp Ann. Manufact. Technol. **59**(2), 628–651 (2010)
5. Cahill, D.G., Ford, W.K., Goodson, K.E., Mahan, G.D., Majumdar, A., Maris, H.J., et al.: Nanoscale thermal transport. J. Appl. Phys. **93**(2), 793–818 (2003)
6. Kolzer, J., Oesterschulze, E., Deboy, G.: Thermal imaging and measurement techniques for electronic materials and devices. Microelectron. Eng. **31**(1–4), 251–270 (1996)
7. Li, Y., Wu, J., Luo, J., Wang, J., Yu, W., Cao, B.: Quantitative measurements in scanning thermal microscopy: theoretical models, calibration technique, and integrated instrument. J. Appl. Phys. **138**(5), 050701 (2025)
8. Yue, Y., Wang, X.: Nanoscale thermal probing. Nano Rev. **3**(1), 11586 (2012)

9. Vetrone, F., Naccache, R., Zamarron, A., Juarranz de la Fuente, A., Sanz-Rodriguez, F., Martinez Maestro, L., et al.: Temperature sensing using fluorescent nanothermometers. Acs Nano. **4**(6), 3254–3258 (2010)
10. de Bastida, G., Arregui, F.J., Goicoechea, J., Matias, I.R.: Quantum dots-based optical fiber temperature sensors fabricated by layer-by-layer. IEEE Sensors J. **6**(6), 1378–1379 (2006)
11. Beechem, T., Graham, S., Kearney, S.P., Phinney, L.M., Serrano, J.R.: Invited article: simultaneous mapping of temperature and stress in microdevices using micro-Raman spectroscopy. Rev. Sci. Instrum. **78**(6), 061301 (2007)
12. Serrano, J.R., Phinney, L.M., Rogers, J.W.: Temperature amplification during laser heating of polycrystalline silicon microcantilevers due to temperature-dependent optical properties. Int. J. Heat Mass Transf. **52**(9–10), 2255–2264 (2009)
13. Dwilinski, R., Doradzinski, R., Garczynski, J., Sierzputowski, L.P., Puchalski, A., Kanbara, Y., et al.: Bulk ammonothermal GaN. J. Cryst. Growth. **311**(10), 3015–3018 (2009)
14. Huang, D., Sun, Q., Liu, Z., Xu, S., Yang, R., Yue, Y.: Ballistic thermal transport at sub-10 nm laser-induced hot spots in GaN crystal. Adv. Sci. (Weinh.). **10**(2), 2204777 (2023)
15. Balkanski, M., Jain, K.P., Beserman, R., Jouanne, M.: Theory of interference distortion of Raman scattering line shapes in semiconductors. Phys. Rev. B. **12**(10), 4328–4337 (1975)
16. Al-Zuhairi, O., Shuhaimi, A., Nayan, N., Azman, A., Kamarudzaman, A., Alobaidi, O., et al.: Non-polar gallium nitride for photodetection applications: a systematic review. Coatings. **12**(2), 275 (2022)
17. Jabbar, H.D., Fakhri, M.A., AbdulRazzaq, M.J.: Gallium nitride–based photodiode: a review. Mater. Today Proc. **42**(5), 2829–2834 (2020)
18. Liu, C., Kang, J.C., Huang, Z.Q., Song, Y.H., Xiao, Y.S., Song, J., et al.: Gallium nitride catalyzed the direct hydrogenation of carbon dioxide to dimethyl ether as primary product. Nat. Commun. **12**(1), 2305 (2021)
19. Hu, Y.J., Zeng, L.P., Minnich, A.J., Dresselhaus, M.S., Chen, G.: Spectral mapping of thermal conductivity through nanoscale ballistic transport. Nat. Nanotechnol. **10**(8), 701–706 (2015)
20. Chou, T.H., Lindsay, L., Maznev, A.A., Gandhi, J.S., Stokes, D.W., Forrest, R.L., et al.: Long mean free paths of room-temperature THz acoustic phonons in a high thermal conductivity material. Phys. Rev. B. **100**(9), 094302 (2019)

Chapter 8
Raman Characterization of Phonon Non-equilibrium

Abstract The performance and reliability of nanoscale devices are strongly influenced by the transport and dissipation of energy from hot charge carriers generated under optical or electrical excitation. In semiconductors, hot carriers relax their excess energy through scattering with phonons, after which energy dissipation proceeds via phonon transport. However, at the nanoscale, limited phonon–phonon interactions and insufficient thermalization can drive phonon populations out of equilibrium, resulting in distinct local temperatures for optical and acoustic phonon modes. This phonon non-equilibrium significantly affects heat dissipation efficiency and can degrade the thermal stability of nanoelectronic systems. This chapter focuses on Raman-based characterization of phonon non-equilibrium in various material systems. Section 8.1 introduces the experimental and theoretical challenges associated with probing non-equilibrium phonon populations. Section 8.2 describes Raman approaches for quantifying optical phonon temperatures. Section 8.3 discusses techniques for distinguishing optical and acoustic phonon temperatures in two-dimensional (2D) materials. Section 8.4 extends these methods to three-dimensional (3D) crystals. Section 8.5 explores phonon non-equilibrium at interfaces, where mismatched vibrational spectra hinder thermalization across material boundaries.

Keywords Phonon non-equilibrium · Optical phonons · Acoustic phonons

8.1 Challenges in Phonon Non-equilibrium Characterization

Phonon non-equilibrium refers to a state where the energy distribution among different phonon modes deviates from the thermal equilibrium. Phonons, as quantized units of lattice vibrations, exist in various forms: optical phonons (OPs) and acoustic phonons (APs). In a state of equilibrium, these phonon modes efficiently exchange energy through processes such as phonon-phonon scattering, ensuring the entire lattice operates at a consistent thermal state. However, under nanoscale

Y. Yue, *Raman Thermometry*, https://doi.org/10.1007/978-981-95-5985-5_8

excitation, whether from focused laser irradiation or localized electrical bias, the energy injected into specific phonon modes can lower their ability to transfer this energy to other modes, as shown in Fig. 8.1a. This non-equilibrium leads to a scenario where OPs may accumulate excess energy and exhibit a higher effective temperature compared to acoustic phonons, as displayed in Fig. 8.1b.

The impact of phonon non-equilibrium to devices is significant. Acoustic phonons, with their relatively high group velocities and substantial contribution to thermal conductivity, are primarily responsible for transporting heat away from localized hotspots in devices. When optical phonons remain in a hot state due to weak coupling with acoustic phonons, a phenomenon observed is that the transfer of energy from the excited region is slowed down. This trapped energy causes localized overheating, as the heat generated by hot carriers cannot be effectively conducted to cooler parts of the device. For downscaled devices, where the volume available for heat spreading is extremely limited, such overheating leads to thermal stress and accelerates material degradation. Thus, a fundamental understanding and experimental characterization of phonon non-equilibrium are critical for the development of nanoelectronics.

Measuring non-equilibrium phonon effects is challenging due to the extreme temporal and spatial scales involved. The transient nature of phonon dynamics, with lifetimes spanning picoseconds to nanoseconds, demands advanced ultrafast techniques such as femtosecond laser pump-probe spectroscopy to resolve rapid thermal and vibrational processes. These methods must not only generate tightly controlled excitation pulses but also capture time-resolved signals with sufficient sensitivity to distinguish between the fleeting energy states of different phonon modes. Additionally, the spatial localization of non-equilibrium phonon distributions confined to nano-to-micrometer regions requires high-resolution tools to capture localized thermal transport.

We introduced the scanning thermal microscopy (SThM) [3] in above sections. It is equipped with a nanoscale tip, and enables direct spatial mapping of thermal transport parameters (e.g., thermal conductivity, local temperature) with spatial resolution down to sub-100 nm. As a contact-based technique, SThM relies on heat flow to maintain the tip at a constant temperature, a process directly influenced by

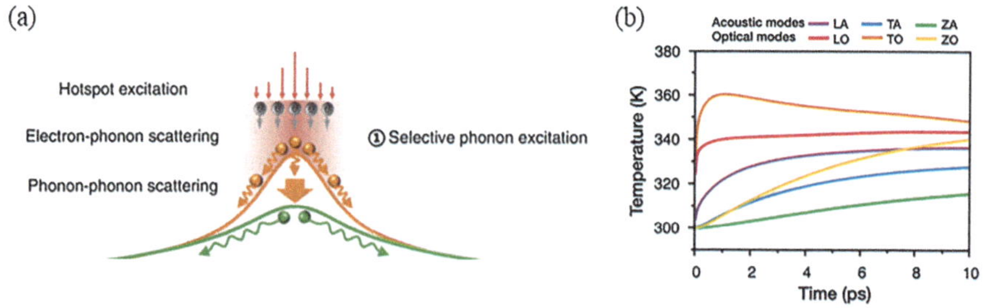

Fig. 8.1 (**a**) Origins of phonon non-equilibrium and (**b**) the resulting phenomena. (Reprinted from ref. [1, 2], with permission from American Physical Society)

the sample's thermal properties and surface topography. However, the complex heat transfer mechanisms at the tip-sample interface make it difficult to accurately estimate tip-sample interfacial thermal resistance. As such, the development of experimental approaches to characterize phonon non-equilibrium is essential for both fundamental understanding and practical applications.

8.2 Raman Characterization of Optical Phonon Temperatures

Raman thermometry offers unique advantages over traditional thermal probing methods, particularly at the micro/nanoscale, due to its non-contact nature and ability to access tiny spatial domains [4]. In contrast, common contact-mode techniques, such as thermocouples and thermal resistors, require physical attachment of a probe to the sample surface, with temperature determined based on voltage changes across the resistor. Thermal paste is often used to ensure reliable thermal contact, but this inevitably introduces sample contamination or physical damage, limiting applicability to delicate nanomaterials [5]. Non-contact optical temperature measurement methods like Raman thermometry avoid these issues, making them more versatile for a wider range of applications.

Beyond temperature measurement, Raman spectroscopy can distinguish optical phonon modes based on the atomic displacement patterns of lattice vibrations, shown in Fig. 8.2. For instance, molybdenum disulfide (MoS_2) exhibits two sets of Raman modes: out-of-layer modes, associated with flexural optical (ZO) phonons, and in-plane modes, linked to longitudinal/transverse optical (LO/TO) phonons [6]. Graphene features distinct G-band (near 1580 cm^{-1}) and 2D-band (near 2670 cm^{-1}) peaks [7]: the G-band arises from the emission or absorption of a zone-center LO phonon by an excited photon, while the 2D-band involves second-order, double-resonant processes with two in-plane TO phonons of opposite and relatively large wavevectors. In gallium nitride (GaN) samples, two active modes are observable,

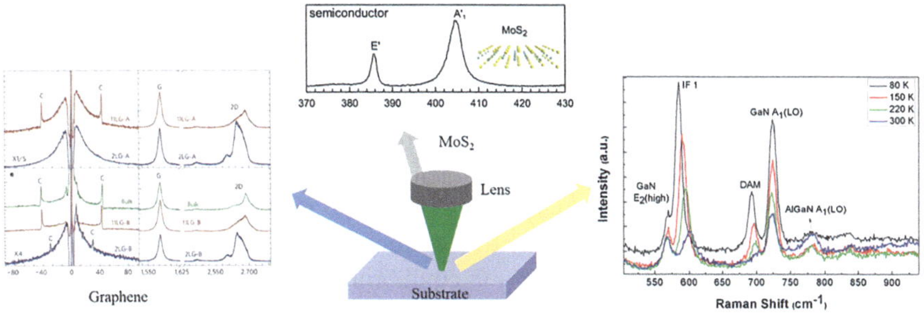

Fig. 8.2 Raman characterization of optical phonon modes (Reprinted from ref. [6, 7, 9], with permission from Springer and American Physical Society)

the E_2(TO) and A_1(LO) peaks, representing transverse and longitudinal optical phonon modes at the center of the Brillouin zone [8]. This capability highlights the value of temperature-dependent Raman, which enables selective probing of OP temperatures and facilitates the characterization of phonon non-equilibrium.

For single-layer graphene (SLG), first-principles modelling shows that thermal conductivity measurement with Raman can be overestimated if not considering the non-equilibrium effect [10]. By linking G-band and 2D-band Raman responses to OP and intermediate-frequency phonon temperatures, they revealed that flexural acoustic (ZA) phonons, primary heat carriers in SLG, exhibit the strongest non-equilibrium with OPs due to weak inter-phonon coupling. This is because Raman signals track only optical phonon temperatures, while the ZA modes that dominate thermal transport stay cooler. Experimental work [11] further expanded Raman's utility by analysing both first-order (G-band, 2D-band) and second-order (D + D″, 2D′) peaks in suspended graphene to resolve temperatures of specific AP polarizations. Anti-Stokes/Stokes intensity ratios of the G-band provided OP temperatures, while shifts in the D + D″ peak, sensitive to ZA phonons, revealed significantly lower ZA temperatures. This mode-specific resolution demonstrated that Raman can distinguish non-equilibrium not just between OPs and APs, but also among AP polarizations, as different peaks couple to distinct phonon populations. It is noted that the laser spot size amplifies this effect: submicron spots restrict phonon thermalization, enhancing temperature differences that Raman detects.

Recently, an experimental work [12] utilized G-band Raman shifts to track OP temperatures in graphene paper (GP), correlating these shifts with independently measured thermal conductivity and laser absorption data to derive AP temperatures. This approach directly quantified OP-AP non-equilibrium: under ~1 μm laser spot heating, OP temperature rises were over 80% higher than those of APs, and the energy coupling factor between in-plane OPs and APs was (1.59–3.10) × 10^{15} W/(m^3K), aligning with quantum mechanical predictions. The study underscored that Raman's non-contact nature avoids sample contamination, a critical advantage over contact-based thermometry that could disrupt delicate phonon thermal states.

In addition, they extended these findings to supported two-dimensional (2D) materials, using Raman spectroscopy to characterize OP-AP non-equilibrium in monolayer WSe_2 on SiO_2, a system where interfacial heat transport complicates phonon dynamics [13]. By analysing shifts in the degenerate 260 cm^{-1} peak (probing OPs) and correlating them with frequency-modulated Raman data (probing AP-driven interfacial transport), they found that OP-AP temperature differences (ΔT_{OA}) accounted for up to 48% of the total OP temperature rise under small laser spots (~0.6 μm). This pronounced non-equilibrium arose from enhanced phonon scattering at the WSe_2-SiO_2 interface, which Raman captured via changes in peak shift magnitudes with laser spot size. The study also used Raman-derived ΔT_{OA} to correct interfacial thermal resistance measurements, emphasizing that Raman is indispensable for isolating phonon non-equilibrium from other thermal effects in supported systems. These studies highlight Raman spectroscopy as a powerful probe of phonon nonequilibrium. It enables non-contact measurement of OP temperatures through mode-specific peak shifts, distinguishes the thermal responses of OP and AP modes. In both suspended and supported 2D materials, Raman spectroscopy provides direct access to micro/nanoscale thermal gradients and mode-specific temperatures, establishing it as an indispensable tool for understanding phonon nonequilibrium in nanodevice design.

8.3 Distinguishing Optical/Acoustic Phonon Temperatures in 2D Materials

As mentioned before, Raman spectroscopy has been widely used to measure the thermal properties of 2D materials. Regardless of the different Raman-based methods employed, the underlying physical process is similar: it involves energy transfer among carriers including photons, electrons, and phonons. However, this inter-phonon branch thermal non-equilibrium has less been considered in thermal conductivity measurement of 2D materials.

The ns ET-Raman technique can be used to explore the temperature non-equilibrium among different phonon branches. Figure 8.3a illustrates the energy transfer process among different energy carriers, providing an overview of the physical process. Following laser irradiation, electrons absorb photon energy and are excited to generate electron-hole pairs. These hot carriers then diffuse and recombine, releasing energy by scattering with OP. For a suspended 2D material, it is physically reasonable to assume that the absorbed photon energy is transferred from hot carriers to OPs.

OPs receive energy from hot carriers, leading to a prominent temperature rise. Each OP branch will have different energy transfer to AP branches. After receiving energy from different OPs, APs transfer the energy to the edge of the suspended 2D material by heat conduction. However, this heat conduction is different for the three acoustic phonon branches. As shown in Fig. 8.3b, the temperature difference between OP and AP (ΔT_{OA}) and temperature rise of APs (ΔT_{AP}) vary differently with laser spot size. In the Raman-based temperature measurement, peaks correspond to OP branches, therefore, the probed temperature rise (ΔT_{m}) reflects OP temperature. For large enough laser spots, this temperature approximates AP temperature, and the determined thermal conductivity can be considered intrinsic to the sample. As shown in Fig. 8.3c, ΔT_{m} can be measured under heating with different laser spot size. Fitting the $\Delta T_m \sim r_0$ data with a specific function allows quantification of the contributions of ΔT_{OA} and ΔT_{AP} to the total probed temperature rise.

Figure 8.4a shows the variation of $\Delta \overline{T}_{\mathrm{AP}}\big|\mathrm{CW}$ with laser spot size in modelling for the 55 nm thick MoS_2. This relation reflects the Raman-probed temperature rise and how acoustic phonon temperature is affected by spot size. Figure 8.4b shows the Raman shift power coefficient (Ψcw) values and fitting curves for two Raman modes under three objective lenses. The A values, mentioned previously in the fitting for $\Delta T_m \sim r_0$ data, for the E_{2g}^1 and A_{1g} modes are determined. At 532 nm, the MoS_2 has

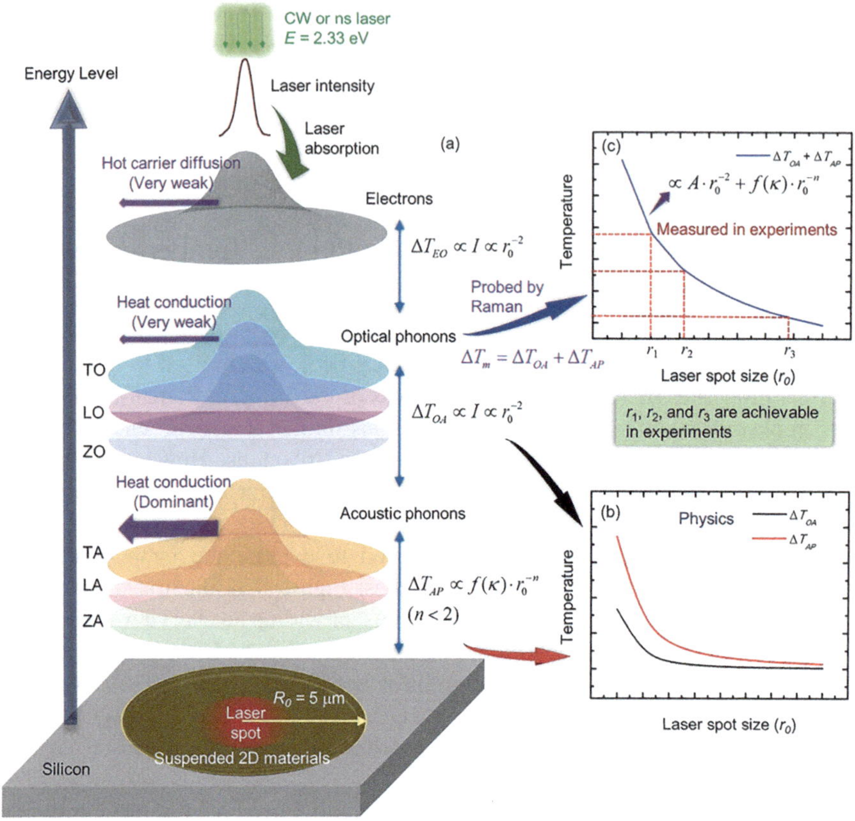

Fig. 8.3 (**a**) Schematic of CW or ns laser irradiation on suspended 2D material. (**b**) Temperature difference between optical and acoustic phonons. (**c**) Sample thermal conductivity and the optical-acoustic phonon coupling coefficient [14]

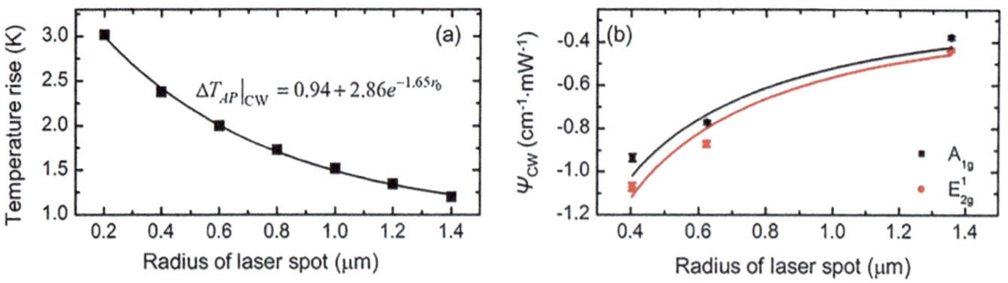

Fig. 8.4 (**a**) Theoretical temperature rise of acoustic phonons in MoS_2 under CW laser heating. (**b**) ψ_{CW} against laser spot size for MoS_2 [14]

an estimated laser absorption of 40.5%. From these A values, the Raman temperature coefficients of MoS_2 are estimated as −0.016 and −0.0171 cm^{-1}/K for the E^1_{2g} and A_{1g} modes. After fitting, the OP-AP energy coupling factors (Gpp|CW) for E^1_{2g} and A_{1g} modes under CW laser are further determined.

As shown in Fig. 8.5a, LO/TO phonon temperatures are slightly higher than ZO phonon temperatures, and both are much higher than AP temperatures. However, this temperature difference decreases with increasing laser spot size, meaning the contribution of ΔT_{OA} to the Raman-measured temperature rise diminishes, shown in Fig. 8.5b. Figure 8.4c also shows that ΔT_{OA} contributions from the two optical phonon branches are very close under CW laser, consistent with their Gpp|CW values. Importantly, for small laser spots, the OP-AP temperature difference accounts for >25% of the measured temperature rise. Thus, neglecting this difference in Raman measurements with small laser spots would significantly underestimate thermal conductivity. These results also show that the energy exchange between different OPs is negligible due to their small temperature difference compared to that between OPs and Aps, which has a much larger temperature difference, as illustrated in Fig. 8.5a.

Figure 8.5b shows distinguished temperatures of LO/TO phonon, ZO phonon, and APs in MoS_2 under ns laser heating. Similar to the CW case, LO/TO phonons

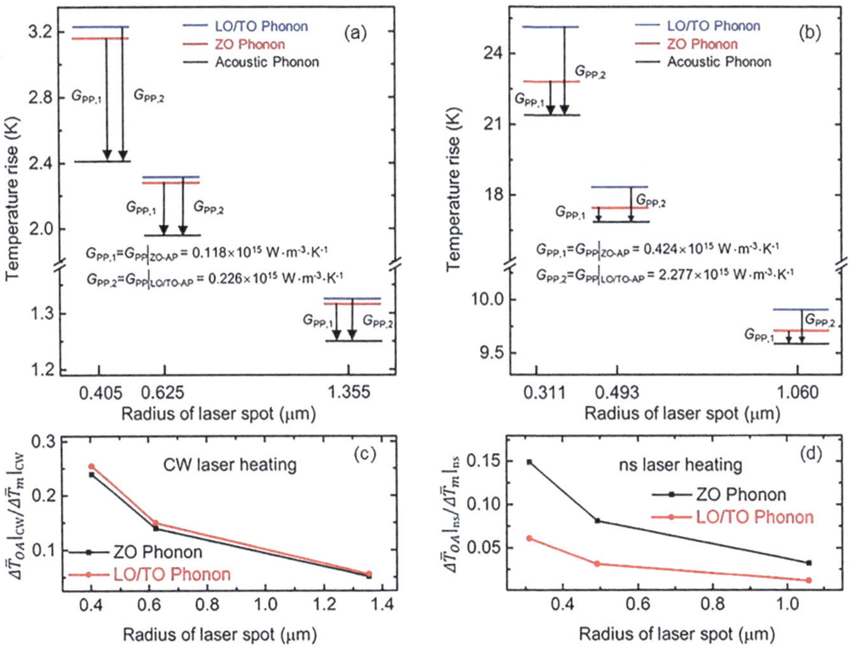

Fig. 8.5 Temperatures of LO/TO, ZO phonons, and APs under (**a**) CW and (**b**) ns laser heating. Phonon contributions of LO/TO and ZO modes to the overall temperature response under (**c**) CW and (**d**) ns laser heating [14]

have the highest temperature during laser heating. Under ns laser, *G*pp|ns of LO/TO phonon is much larger than that of ZO phonon, so the ZO phonon/AP temperature difference is much smaller than that of LO/TO phonon/AP. Comparing ΔT_{OA} contributions under CW and ns lasers, the contribution under CW laser is much larger, as shown in Fig. 8.5c, d. This arises from two effects. First, under same laser intensity, ns laser heating does not reach steady state, leading to a lower overall temperature rise (ΔT_{m}) and a lower ΔT_{OA} due to its higher *G*pp. Second, ΔT_{OA} is reduced more than ΔT_{m} under ns laser compared to CW laser.

For $MoSe_2$, the in-plane thermal conductivity of the sample ($\kappa_{\parallel}$) obtained under 20× objective lens is 14.6 ± 0.6 W/(m·K). Previous studies have also measured $\kappa_{\parallel}$ of $MoSe_2$ of different thickness. There is an increasing trend for $\kappa_{\parallel}$ with the increased thickness due to surface phonon scattering. Using the same process as for MoS_2 case, the ZO phonon and acoustic phonon temperature under CW and ns laser heating is obtained. The temperature difference between ZO phonon and AP decreases with the increased laser spot size. The percentage contribution of ΔT_{OA} is decreasing with the increased laser spot size. The contribution of ΔT_{OA} under CW laser decreases from around 32% to around 6% when the laser spot size increases from 0.405 μm to 1.355 μm. While the contribution of ΔT_{OA} under ns laser decreases from around 20% to around 5% when the laser spot size increases from 0.311 μm to 1.060 μm. That is, the contribution of ΔT_{OA} under CW laser is larger than the contribution of ΔT_{OA} under ns laser.

In this work, a breakthrough is made in distinguishing and measuring the temperatures of OPs and APs for these materials. Compared to *G*pp for MoS_2, $MoSe_2$ has much smaller *G*pp values, reflecting weaker OP-AP coupling. For both materials, OP-AP energy coupling factors are weaker under CW laser heating than ns laser heating. This difference may originate from the higher electron population or stronger spatial thermal non-equilibrium induced by ns laser heating, although further studies are required for a detailed explanation. Readers can refer to the reference [14] for more details about this work.

8.4 Probing Phonon Non-equilibrium Effect in 3D Crystals

Raman spectroscopy enables direct measurements of this phonon temperature non-equilibrium via the Raman spectrum-temperature dependence. For example, experiments based on Raman spectroscopy have shown that weak coupling between in-plane and out-of-plane phonons in graphene and 2D transition metal dichalcogenides easily induces measurable phonon non-equilibrium, even near microscale laser-heated hotspots. However, phonon non-equilibrium has never been explicitly demonstrated in realistic three-dimensional (3D) device semiconductors for devices. This gap arises because exciting nanoscale hotspots to induce observable phonon non-equilibrium and characterizing the underlying phonon dynamics are highly challenging, particularly for the sub-10 nm hotspots common in modern nanoelectronics. Hence, experimental evidence and comprehensive understandings of phonon non-equilibrium in 3D electronic and optoelectronic materials remain limited.

GaN is a wide-bandgap semiconductor, and is characterized by high lattice thermal conductivity and a broad phonon mean free path [15]. At nanoscale, when the characteristic length is smaller than the phonon mean free path, GaN exhibits strong phonon non-equilibrium. This is investigated by using a tip-enhanced Raman system that combines localized heating with nanoscale temperature measurement. As illustrated in Fig. 8.6a, a single-crystal GaN sample is placed on an atomic force microscope (AFM) stage and probed with a gold-coated silicon nanotip operated in contact mode to excite an enhanced optical field (see Chap. 7) [15]. Figure 8.6b shows a SEM image of the silicon nanotip used as the probe. Focusing the Raman laser on the tip apex creates a strong optical field at the tip-substrate interface, producing a localized heating region in the GaN subsurface.

In above configuration, GaN exhibits two active modes: the E_2(TO) and $A1$(LO) peaks, representing transverse and longitudinal optical phonon modes at the Brillouin zone center [8]. Phonon temperature rise in the heated region is extracted from Raman peak shifts. Figure 8.6c shows Raman spectra for the E_2(TO) mode at different laser powers. Possible nonlinear deviations in Raman shifts from SERS are suppressed by low excitation power. The small laser incidence angle and the weak absorption of GaN at 532 nm further minimize substrate heating, verified by control experiments without the nanotip, where Raman-derived temperature rises remain below 1 K. The gold coating on the tip also reduces heating and thermal stress, due to its high reflectivity, shallow skin depth, and excellent thermal conductivity.

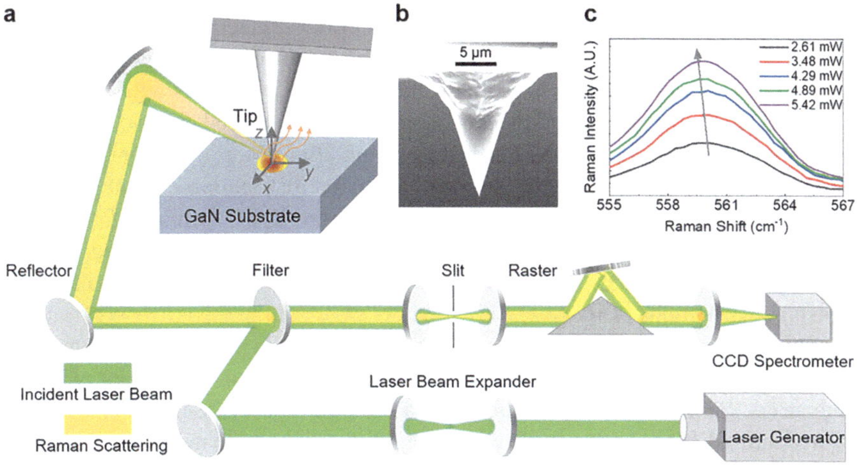

Fig. 8.6 (**a**) Schematic of the experimental setup for tip-enhanced Raman thermal measurement. (**b**) The SEM image of the gold-coated silicon nanotip. (**c**) Raman spectra (E_2(TO) mode) of GaN under different laser excitation [16]

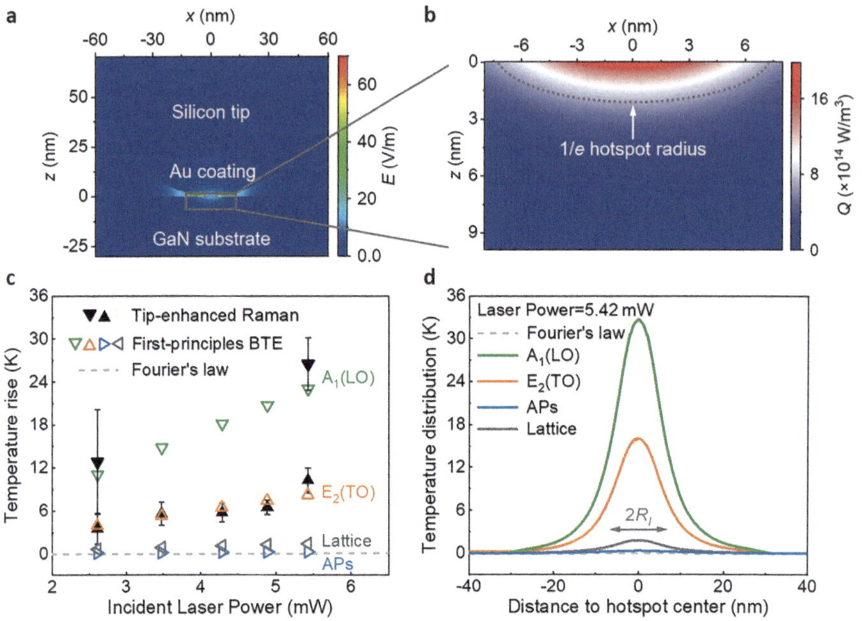

Fig. 8.7 (**a**) Electric field distribution at the tip–substrate interface. (**b**) Heat generation in the GaN substrate. (**c**) Temperature rises of E_2(TO) and A_1(LO) phonons. (**d**) Phonon temperature distribution near the hotspot from first-principles BTE calculations [16]

To interpret experimental results, the nanoscale heated region was characterized via electromagnetic simulations of the tip-substrate system, determining the heat generation rate in GaN substrate. Figure 8.7a shows the electric field distribution at the tip-substrate contact region, revealing that the enhanced electric field is confined to a sub-10 nm radius region beneath the nanotip, with intensity amplified by a factor of 79 due to the gold-coated silicon nanotip. This enhanced electric field creates nanoscale hotspots in the GaN: for the maximum laser power of 5.42 mW, the incident intensity is 1.9×10^6 W/m^2 and the calculated distribution of heat generation rate distribution in Fig. 8.7b shows a hotspot radii are $R_l = 7.2$ nm and $R_v = 2.1$ nm, respectively. Determining exact heat generation at this scale is challenging, but electromagnetic simulations are a well-established method for calculating nanoscale hotspot distributions [15].

The results based on first-principles phonon Boltzmann transport equation (BTE) calculations are presented in Fig. 8.7c, showing excellent agreement with tip-enhanced Raman thermal measurements across a range of laser powers. Distinct temperature differences are observed among different phonon modes, including the A_1(LO) (lower triangles) and the E_2(TO) (upper triangles) phonon modes, the average temperature of acoustic phonons (left triangles), and the lattice temperature (right triangles). These variations provide direct evidence of pronounced phonon non-equilibrium near the nanoscale hotspot. In particular, the temperature rises of the A_1(LO) and E_2(TO) modes are substantially higher than those predicted either by Fourier's law at the nanoscale or by the acoustic phonon response. The consistency between measurements and theoretical predictions therefore confirms the existence of nanoscale phonon non-equilibrium in GaN.

To further elucidate this phenomenon, Fig. 8.7d shows the calculated phonon temperature distribution near the laser-excited hotspot on the GaN surface in contact with the nanotip at a laser power of 5.42 mW. Significant temperature rises of optical phonons are observed, whereas the average acoustic phonon temperature increase is nearly two orders of magnitude smaller. Moreover, a clear non-equilibrium is evident even among different optical phonon branches, providing further validation of this effect.

To experimentally and theoretically establishes unified evidence of phonon non-equilibrium near nanoscale hotspots in three-dimensional semiconductor, a tip-enhanced Raman thermal measurement is conducted. Quantitative agreement with first-principles-based phonon BTE calculations further validates these findings. The unusually strong phonon non-equilibrium is shown to originate from highly selective phonon excitation via strong Fröhlich electron–phonon coupling, combined with weak optical–acoustic phonon coupling arising from the large AO frequency gap in GaN. Moreover, this phenomenon is demonstrated to be universal in three-dimensional materials at the nanoscale, particularly in III-V polar semiconductors. The approach presented here provides a viable route for probing phonon non-equilibrium at the nanoscale, offering new insights into the complex non-equilibrium dynamics of energy carriers in nanoelectronics.

8.5 Phonon Non-equilibrium at the Interface

Efficient interfacial heat transfer and fast dissipation are thus crucial for reliable device operation. Recent studies reveal that interfacial phonon modes play a key role in cross-interface thermal transport, and recent theories provide new insights into how specific phonon modes contribute to this process [17, 18]. Yet experimental studies remain limited. Here, Raman is employed to probe phonon non-equilibrium at interfaces in 2D materials.

Figure 8.8a shows the layered atomic structure of MoS_2 and Fig. 8.8b presents the typical Raman spectra of the MoS_2/c-Si sample under 532 nm laser excitation. MoS_2 exhibits two dominant peaks: the E_{2g}^1 mode near 385 cm^{-1} and the A_{1g} mode near 410 cm^{-1}, along with the longitudinal optical phonon mode of c-Si at ~522 cm^{-1}. The E_{2g}^1 mode corresponds to the in-plane vibration of Mo and S atoms in opposite directions, associated with longitudinal/transverse optical phonons, while the A_{1g} mode arises from the out-of-plane vibration of S atoms, linked to flexural optical phonons [20]. Thus, the Raman shifts of E_{2g}^1 and A_{1g} mode are employed to distinguish the contributions of in-plane and out-of-plane optical phonons to interfacial thermal transport.

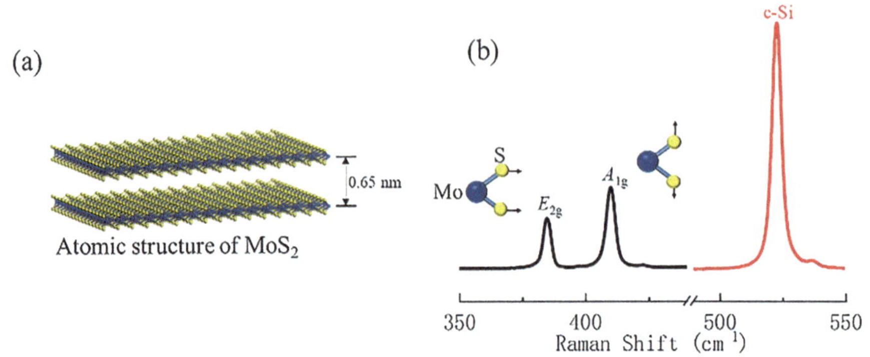

Fig. 8.8 (**a**) Side view of the atomic structure of the MoS_2 film. (**b**) Raman spectra of the MoS_2/c-Si sample under 532 nm laser excitation, used to determine the temperature rise of the sample. (Reprinted from ref. [19], with permission from AIP Publishing)

Temperature-dependent Raman spectroscopy was used to determine temperature rises under laser heating. The extracted linear temperature coefficients are −0.0135, −0.0131, and −0.0218 cm^{-1}/K for the E^1_{2g}, A_{1g}, and Si modes of the MoS_2/c-Si sample, respectively. Power-dependent Raman spectra were then measured using 50× and 100× objectives, as shown in Fig. 8.9a, b. Under 50× objective, the first-order power-dependent coefficient (χ_P) is −0.0298 and − 0.0301 cm^{-1}/mW for the E^1_{2g} and A_{1g} modes, respectively. Under 100× objective lens, χ_P values are −0.0930 and − 0.0875 cm^{-1}/mW for the E^1_{2g} and A_{1g} modes, respectively. Redshifts of the Si mode confirm substrate temperature rises in both cases. The average temperature difference ($\Delta\overline{T}$) between the MoS_2 film and c-Si substrate was calculated as 2.02 K (E^1_{2g}) and 2.21 K (A_{1g}) under the 50× objective, and 6.61 K (E^1_{2g}) and 6.36 K (A_{1g}) under the 100× objective, highlighting more pronounced interfacial resistance under tighter laser focusing.

3D finite-volume simulations were conducted to determine the interfacial thermal conductance (G) and in-plane conductivity (k_s) of MoS_2 on c-Si. The model used experimental geometry, cross-plane conductivities of 2 W/mK (MoS_2) [21] and 148 W/mK (c-Si), and 1 mW laser heating. As shown in Fig. 8.9c, multiple reflections at the MoS_2/c-Si interface were included via Transfer Matrix Method (TMM) [22] analysis of optical absorption. In modeling, temperature differences between the MoS_2 film and c-Si substrate under 50× and 100× objectives were calculated for various G and k_S. The measured temperature rises represent Raman-intensity–weighted averages, with penetration depths of 38.5 nm for MoS_2 and 820 nm for c-Si. Thus, the film and substrate temperature rises were obtained by weighted integration of $\Delta\overline{T}$. The true G and k_S were identified when calculated temperature differences matched both experimental conditions. Results show that lower G and k_S yields higher temperature rises. Reduced k_S hinders lateral heat spreading, while reduced G requires a larger local rise to transfer heat into c-Si. Under 50×, $\Delta\overline{T}$ is more sensitive to G, whereas under 100×, $\Delta\overline{T}$ depends more strongly on k_S,

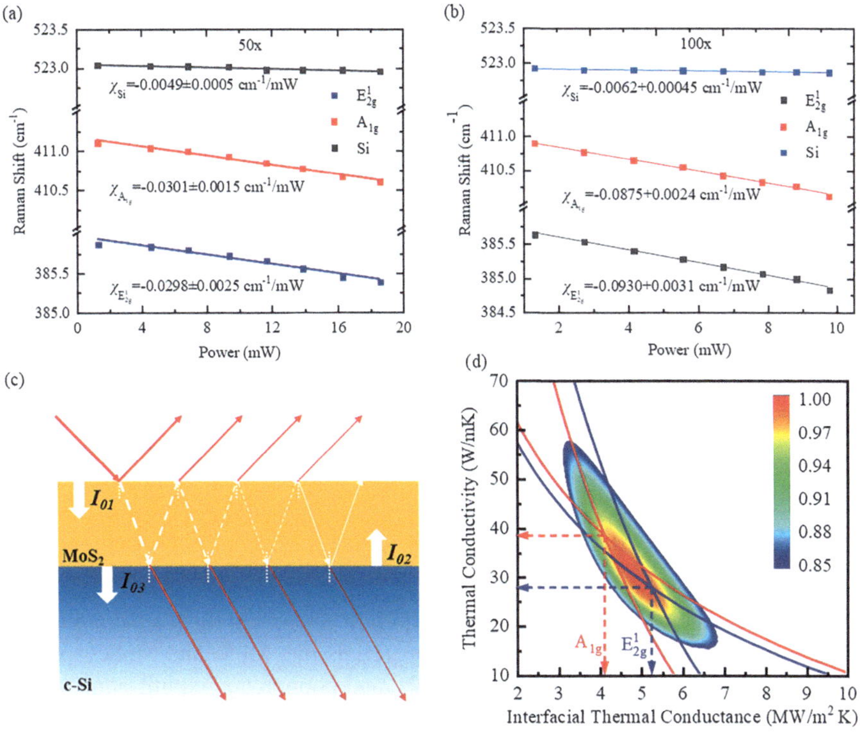

Fig. 8.9 (**a**) Schematic of laser transmission. Raman intensity–weighted average temperature differences for a series of G and k_S values under (**b**) 50× and (**c**) 100× objectives. (**d**) Extracted G and k_S values using the E^1_{2g} and A_{1g} modes, with uncertainty region indicated. (Reprinted from ref. [19], with permission from AIP Publishing)

especially at low G. This explains the different power-dependent coefficients of the E^1_{2g} and A_{1g} modes under the two objectives.

For each laser heated spot size, the experimentally obtained temperature difference between MoS_2 film and c-Si substrate (the isolines) could be satisfied by multiple G–k_S combinations. To resolve this, results from both objectives and a normalized probability distribution function Ω were used. Ω is defined as $\Omega = \exp\left[-\left(x-\overline{x}\right)^2/\left(2\sigma^2\right)\right]$, where x, $\overline{x}$, and σ are the temperature difference, its mean and standard deviation, respectively. A composite probability function, $\Omega_{(G,k_S)} = \Omega_{\Delta\overline{T}_{50x}} \cdot \Omega_{\Delta\overline{T}_{100x}}$ was applied in the G–k_S space. The values of G and k_S were determined from the intersection of the 50× and 100× curves, with uncertainties indicated by the $\Omega_{(G,k_s)} = 0.6065$ contour. The obtained G by E^1_{2g} and A_{1g} modes are $5.26^{+1.27}_{-0.75}$ and $4.13^{+1.46}_{-0.96}$ mW/(m^2·K), respectively. The G from the out-of-plane A_{1g} mode was ~21.5% lower than that from the in-planeE^1_{2g} mode, consistent with observations in MoS_2/quartz interfaces. This anisotropy arises from different acoustic phonon branches exhibiting distinct interfacial resistances, leading to mode-dependent temperature rises and G values. Overall, the results indicate superior cross-interface transport via in-plane phonon modes in supported 2D materials.

To sum up, the interfacial thermal transport behavior of MoS_2 films on c-Si substrates was examined using the steady-state Raman technique. Both the E^1_{2g} and A_{1g} vibrational modes of MoS_2 display a linear softening with increasing temperature. By analyzing the thermal response of the MoS_2/c-Si system under two laser spot sizes and incorporating 3D simulations, the interfacial thermal conductance (G) and in-plane thermal conductivity (k_S) for each mode were extracted. The extracted G using A_{1g} mode is approximately 21.5% lower than the E^1_{2g} mode. Besides, the thermal transport difference between the two phonon modes is also reflected in k_S. These findings highlight the significant role of phonon mode dependence in the thermal transport of 2D materials on substrates, offering deeper insights into heat conduction mechanisms in supported two-dimensional systems.

8.6 Summary

This chapter focuses on phonon non-equilibrium, including distinguishing optical and acoustic phonon temperatures in 2D materials like MoS_2 and $MoSe_2$, phonon non-equilibrium in 3D crystals such as GaN, and phonon non-equilibrium at interfaces in 2D materials like MoS_2/c-Si. Future studies could enhance Raman temporal resolution to probe ultrafast phonon dynamics and monitor non-equilibrium states during device operation. Exploring phonon non-equilibrium in more complex heterostructures and under extreme conditions (e.g., high pressure or strong electric fields) may reveal new thermal transport mechanisms. Integrating these findings with device design could lead to more efficient thermal management strategies for high-performance nanoelectronics.

References

1. Xu, J., Hu, Y., Bao, H.: Quantitative analysis of nonequilibrium phonon transport near a nanoscale hotspot. Phys. Rev. Appl. **19**(1), 014007 (2023)
2. Tong, X., Bernardi, M.: Toward precise simulations of the coupled ultrafast dynamics of electrons and atomic vibrations in materials. Phys. Rev. Res. **3**(2), 023072 (2021)
3. Li, Y., Wu, J., Luo, J., Wang, J., Yu, W., Cao, B.: Quantitative measurements in scanning thermal microscopy: theoretical models, calibration technique, and integrated instrument. J. Appl. Phys. **138**(5), 050701 (2025)
4. Childs, P.R., Greenwood, J., Long, C.: Review of temperature measurement. Rev. Sci. Instrum. **71**(8), 2959–2978 (2000)
5. Yue, Y., Zhang, J., Wang, X.: Micro/nanoscale spatial resolution temperature probing for the interfacial thermal characterization of epitaxial graphene on 4H-SiC. Small. **7**(23), 3324–3333 (2011)

6. Cong, X., Liu, X.-L., Lin, M.-L., Tan, P.-H.: application of Raman spectroscopy to probe fundamental properties of two-dimensional materials. NPJ 2D Mater. Appl. **4**(1), 13 (2020)
7. Tan, P.-H., Han, W.P., Zhao, W.J., Wu, Z.H., Chang, K., Wang, H., Wang, Y.F., Bonini, N., Marzari, N., Pugno, N.: The shear mode of multilayer graphene. Nat. Mater. **11**(4), 294–300 (2012)
8. Davydov, V.Y., Kitaev, Y.E., Goncharuk, I.N., Smirnov, A.N., Graul, J., Semchinova, O., Uffmann, D., Smirnov, M.B., Mirgorodsky, A.P., Evarestov, R.A.: Phonon dispersion and Raman scattering in hexagonal GaN and AlN. Phys. Rev. B. **58**(19), 12899–12907 (1998)
9. Liu, Y., Chen, D., Wei, G., Lin, Z., He, A., Li, M., Wang, P., Zhang, R., Zheng, Y.: Temperature-dependent ultraviolet Raman scattering and anomalous Raman phenomenon of AlGaN/GaN heterostructure. Opt. Express. **27**(4), 4781–4788 (2019)
10. Vallabhaneni, A.K., Singh, D., Bao, H., Murthy, J., Ruan, X.L.: Reliability of Raman measurements of thermal conductivity of single-layer graphene due to selective electron-phonon coupling: a first-principles study. Phys. Rev. B. **93**(12), 125432 (2016)
11. Sullivan, S., Vallabhaneni, A., Kholmanov, I., Ruan, X.L., Murthy, J., Shi, L.: Optical generation and detection of local nonequilibrium phonons in suspended graphene. Nano Lett. **17**(3), 2049–2056 (2017)
12. Zobeiri, H., Hunter, N., Wang, R.D., Wang, T.Y., Wang, X.W.: Direct characterization of thermal nonequilibrium between optical and acoustic phonons in graphene paper under photon excitation. Adv. Sci. **8**(12), 2004712 (2021)
13. Hunter, N., Azam, N., Zobeiri, H., Van Velson, N., Mahjouri-Samani, M., Wang, X.W.: Interface thermal resistance between monolayer WSe2 and SiO2: Raman probing with consideration of optical-acoustic phonon nonequilibrium. Adv. Mater. Interfaces. **9**(7), 2102059 (2022)
14. Wang, R.D., Zobeiri, H., Xie, Y.S., Wang, X.W., Zhang, X., Yue, Y.N.: Distinguishing optical and acoustic phonon temperatures and their energy coupling factor under photon excitation in nm 2D materials. Adv. Sci. **7**(13), 2000097 (2020)
15. Huang, D.Z., Sun, Q.S., Liu, Z.Y., Xu, S., Yang, R.G., Yue, Y.A.: Ballistic thermal transport at sub-10 nm laser-induced hot spots in GaN crystal. Adv. Sci. **10**(2), 2204777 (2023)
16. Xu, J., Huang, X., Sheng, Y., Sun, Q., Zhang, H., Bao, H., Yue, Y.: Direct observation of substantial phonon nonequilibrium near nanoscale hotspots in gallium nitride. Adv. Sci. **12**(11), 2411040 (2025)
17. Zhou, Y., Hu, M.: Full quantification of frequency-dependent interfacial thermal conductance contributed by two-and three-phonon scattering processes from nonequilibrium molecular dynamics simulations. Phys. Rev. B. **95**(11), 115313 (2017)
18. Cheng, Z., Li, R., Yan, X., Jernigan, G., Shi, J., Liao, M.E., Hines, N.J., Gadre, C.A., Idrobo, J.C., Lee, E.: Experimental observation of localized interfacial phonon modes. Nat. Commun. **12**(1), 6901 (2021)
19. Li, D., Huang, X., Liu, Z., Liu, W., Xu, S., Yue, Y.: Divergent interfacial thermal transport in MoS2/Si heterostructure over optical phonon modes. Appl. Phys. Lett. **123**(25), 252201 (2023)
20. Taube, A., Judek, J., Łapińska, A., Zdrojek, M.: Temperature-dependent thermal properties of supported MoS2 monolayers. ACS Appl. Mater. Interfaces. **7**(9), 5061–5065 (2015)
21. Liu, J., Choi, G.-M., Cahill, D.G.: Measurement of the anisotropic thermal conductivity of molybdenum disulfide by the time-resolved magneto-optic Kerr effect. J. Appl. Phys. **116**(23), 233107 (2014)
22. Yuan, P.Y., Li, C., Xu, S., Liu, J., Wang, X.W.: Interfacial thermal conductance between few to tens of layered-MoS2 and c-Si: effect of MoS2 thickness. Acta Mater. **122**, 152–165 (2017)

Chapter 9
Raman Characterization of Near-field Heat Transfer

Abstract Near-field heat transfer refers to the enhanced thermal energy exchange between two objects separated by subwavelength gaps, where the dominant mechanisms of heat transfer differ fundamentally from those in the far-field (Liu et al., Nanoscale and Nanoscale Microscale Thermophys Eng. 19:98–126, 2015). This chapter focuses on Raman-based characterization of near-field thermal transport. Section 9.1 provides an introduction to the principles and significance of near-field heat transfer. Section 9.2 outlines the experimental challenges and limitations associated with thermal measurements in the near-field regime. Section 9.3 discusses the implementation of Raman thermometry for probing near-field heat transfer. Section 9.4 examines heat transfer between a nanotip and a substrate, highlighting the heat transfer enhancement in tip-substrate system. Section 9.5 considers the influence of air molecules on heat conduction in the near-field, emphasizing the complex interplay between gaseous and radiative contributions at the nanoscale. Together, these discussions demonstrate the potential of Raman thermometry as an effective tool for investigating near-field thermal phenomena and guiding the design of nanoscale thermal devices.

Keywords Near-field · Heat transfer enhancement · Nanotip-substrate thermal transport · Air molecules

9.1 Introduction to Near-field Heat Transfer

Far-field heat transfer occurs when objects are separated by distances much larger than the wavelength of thermal radiation, relying on propagating electromagnetic waves, as shown in Fig. 9.1. This process is constrained by Planck's law, which sets a maximum limit on the amount of radiative energy that can be exchanged between two objects at given temperatures. In this regime, the radiation is isotropic, and the heat flux depends on the fourth power of the temperature difference between the objects, as described by the Stefan-Boltzmann law. In the near-field, however,

Y. Yue, *Raman Thermometry*, https://doi.org/10.1007/978-981-95-5985-5_9

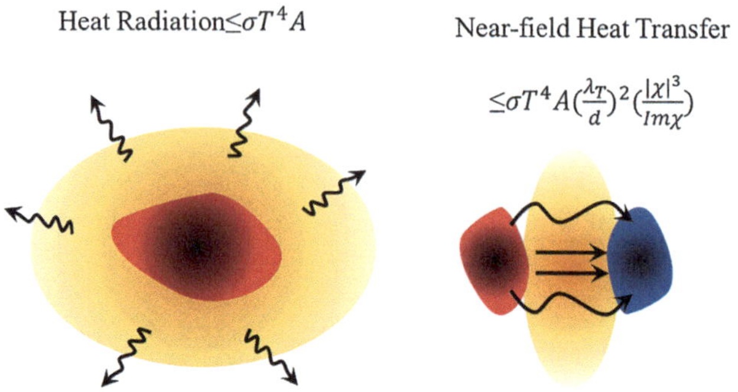

Fig. 9.1 The difference between the far-field and the near-field heat transfer

subwavelength separations render conventional radiative transfer rules inapplicable. Here, evanescent electromagnetic waves, which are non-propagating and decay exponentially with distance, play a dominant role. These waves arise from the fluctuations of charges at the surfaces of the objects and are unable to propagate in free space over long distances, but within the near-field, they can tunnel across the gap, enabling efficient energy transfer. Additionally, surface phonon-polaritons, collective oscillations of phonons and electromagnetic fields at the surface of polar materials such as silicon carbide or hexagonal boron nitride, further amplify this transfer [1]. These polaritons create strong localized electromagnetic fields at the material surfaces, enhancing the coupling between the two objects and facilitating even greater heat exchange.

The theoretical foundation of near-field heat transfer was laid in 1971 [2], when researchers showed that thermal radiation between parallel plates with a nanoscale gap can exceed the blackbody limit by orders of magnitude. This challenged traditional understanding and spurred research into complex geometries, anisotropic materials, and quantum effects at sub-nanometer gaps. Near-field heat transfer is not limited to radiation, as air conduction exhibits distinct behaviour at the nanoscale compared to the macroscale. In the far-field, air conduction occurs through diffusion driven by random molecular motion, but as gaps shrink to nanoscales, it enters transition and ballistic regimes. In the transition regime, molecular mean free paths become comparable to gap size, causing non-diffusive behaviour; in the ballistic regime (gap < mean free path), molecules cross the gap without collisions, increasing conduction efficiency. This combination of enhanced radiation and modified conduction makes near-field heat transfer a complex, multi-mechanism phenomenon.

The significance of near-field heat transfer lies in its potential to revolutionize technologies that rely on precise thermal control at the nanoscale. For example, in nanoscale thermal management, it could enable more efficient heat dissipation from high-power electronic devices, where overheating is a critical limitation. In energy conversion systems, near-field heat transfer can enhance efficiency by optimizing the spectral alignment between thermal emitters and photovoltaic cells. In quantum heat transport, it offers opportunities to explore some phenomena, such as phonon-mediated entanglement between nanoscale objects, which could have implications for quantum information processing. The experimental investigation of near-field heat transfer faces formidable challenges due to its nanoscale nature and the delicate characteristics of the interactions. These challenges span from controlling the nanoscale gap between objects to distinguishing near-field signals from background noise, and they require innovative solutions at the intersection of physics, engineering, and materials science.

9.2 Difficulties in Near-field Thermal Measurements

Near-field heat transfer experiments require exceptional precision. To resolve these effects, two objects must be separated by gaps much smaller than the wavelength of thermal radiation, often on the order of only a few tens of nanometers. At such scales, the measurements become exceptionally sensitive: nanoscale drifts arising from thermal expansion, mechanical vibrations, or instrumental instabilities can significantly alter the gap and thereby the measured heat flux. Compounding these difficulties are surface forces. Van der Waals interactions exert attractive forces that can pull the objects into contact, leading to stiction events that not only compromise the measurement but also irreversibly damage delicate samples.

To address these challenges, researchers employ ultra-precision positioning systems capable of sub-nanometer control, often coupled with active feedback loops that continuously monitor and stabilize the separation. Nevertheless, maintaining a fixed gap over extended periods remains challenging, as ambient temperature fluctuations, acoustic disturbances, and even subtle building vibrations can gradually perturb the system, distorting measurements Heat transfer across the gap involves contributions from near-field radiation, far-field emission, and conduction through air or residual gas. The intrinsic signal is likewise weak: near-field energy exchange involves minute fluxes across nanoscale areas. Consequently, many experiments are conducted under stringent conditions, including vibration isolation, vacuum environments, and thermal shielding. These measurements are technically demanding, costly, and constrain the accessible range of experimental designs and geometries.

Many studies have aimed to quantify near-field heat transfer, particularly the radiative contributions, where accurate temperature measurement is essential. Scanning thermal microscopy (SThM) probe, functioning both as a heater and a thermistor, enables the measurement of thermal conductivity in thin films [3, 4]. It is also used to measure the near-field radiation by integrating an SThM probe with an embedded thermocouple to probe temperature [5]. In addition, radiation enhancement in the near-field regime between two parallel plates has been studied with embedded temperature sensors on/near surfaces [6–8]. These methodologies push forward the near-field heat transfer studies, while it is a struggle to directly measure the temperatures of both heat exchange surfaces. The techniques predominantly rely on single-sided measurements or derive temperatures from thermal resistance and near-surface measurements.

9.3 Raman Thermometry on Near-field Thermal Transport

Raman spectroscopy has emerged as a promising tool to overcome some of these challenges. As a non-contact technique, it enables temperature measurements at the micro- and nanoscale without disturbing the sample or altering the heat transfer dynamics. This capability is particularly valuable for near-field heat transfer experiments, as it enables direct, simultaneous temperature measurements of both objects involved in the heat exchange, something that traditional techniques struggle to achieve. In this section, focus is placed on heat transfer between a nanotip and substrate, which is crucial for extreme manufacturing [9] and high-density data storage [10] realms. A major challenge in near-field heat transfer experiments is maintaining a stable separation between two surfaces at the sub-wavelength scale. In this section, an atomic force microscope (AFM) integrated with a Raman spectrometer is employed to probe near-field heat transport across gaps as small as several tens of nanometers, shown in Fig. 9.2a.

Unlike conventional tip-enhanced approaches, this system adopts a small-angle aligned laser that simultaneously heats the nanotip and excites the Raman response of the substrate. The nanotip position is precisely monitored using optical lever detection, which tracks the deflection of a laser beam reflected from the backside of the cantilever. Fine nanoscale displacement of the Si nanotip is controlled by a piezoelectric actuator driven with calibrated voltages. The fabricated Si nanotip is 17 μm in length, with an apex curvature radius of 10 nm and a half-cone angle of 10° in Fig. 9.2b. The substrate is a bulk 4H-SiC wafer ($10 \times 10 \times 0.5$ mm^3), grown by the seed crystal method and polished to achieve a surface roughness below

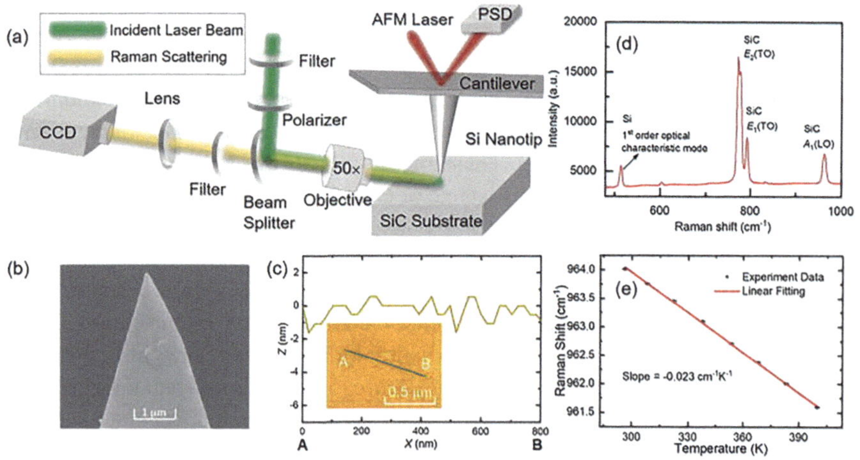

Fig. 9.2 (**a**) Schematic of the experimental setup. (**b**) Scanning electron microscopy (SEM) image of the fabricated Si nanotip. (**c**) AFM image of the SiC substrate. (**d**) Raman spectra acquired from the Si nanotip and SiC substrate. (**e**) Calibration of Raman shift with temperature for the SiC substrate. (Reprinted from ref. [11], with permission from AIP Publishing)

0.9 nm, as displayed in Fig. 9.2c. Raman is then used to capture scattering signals from both the nanotip and the substrate, enabling direct characterization of near-field heat transport.

In our work, a continuous-wave 532 nm laser is tightly focused onto the Si nanotip through a 50× objective lens mounted on a high-precision three-dimensional displacement stage. The incident power is carefully adjusted to ensure effective heat transfer across the tip–substrate interface. Reducing the laser power lowers the temperature rise in both the nanotip and the substrate, thereby weakening interfacial heat transport. At the same time, the diminished excitation also suppresses the Raman scattering intensity, which complicates spectroscopy measurements and increases data-processing uncertainty. The focal spot radius at the sample plane is approximately adjusted as 17.5 μm. Raman scattering signals from the coupled tip-substrate system is collected by the same objective and analysed with a spectrometer equipped with an 1800 grooves/mm grating, providing high spectral resolution.

A typical spectrum is presented in Fig. 9.2d, showing the pronounced Raman modes of the 4H-SiC substrate, along with the first-order optical phonon peak of the Si nanotip [12]. This configuration enables simultaneous in situ thermal probing of both materials. For quantitative analysis, the sharp Raman bands at 520.7 cm^{-1} (Si) and 964.0 cm^{-1} at room temperature, corresponding to A_1(LO) mode of SiC, are selected because of their strong intensities and distinct separation. Peak positions are extracted by Lorentzian fitting, with three spectra averaged for each condition to reduce statistical error. Before acquisition, the system is stabilized for ~30 min to ensure steady-state thermal equilibrium at the interface. Calibration experiments confirm a linear dependence of Raman shift on temperature for both Si and SiC, yielding slopes of −0.022 and −0.023 cm^{-1}/K, as shown in Fig. 9.2e.

9.4 Heat Transfer Analysis Between Nanotip and Substrate

Experiments are performed in ambient air at room temperature. The nanotip is positioned near the substrate, and the objective is aligned to maximize the Si Raman signal at the tip apex. The tip is then adjusted in nanoscale steps (82.5–1320 nm) using a piezo actuator. The total displacement (~1 μm) is small relative to the laser spot (~17.5 μm), ensuring consistent illumination of both tip and substrate. During retraction of the nanotip from the substrate in Fig. 9.3a, b, the Raman shift of the tip shows only marginal variation, whereas the substrate exhibits a pronounced shift, as displayed in Fig. 9.3c, d. Using the linear Raman shift–temperature relationship, the tip temperature remains nearly constant at ~200 °C, while the substrate temperature decreases from ~32.6 to 22.4 °C as the gap increases, shown in Fig. 9.3e.

The focused laser, concentrated by the nanotip onto the SiC substrate, excites Raman scattering from a confined region beneath the tip. This scattered light is reflected by the tip and collected by the objective and CCD, consistent with the observed increase in Si and SiC Raman peak intensities as the tip–substrate gap decreases, as illustrated in Fig. 9.3c, d. Therefore, the Raman signal reflects the

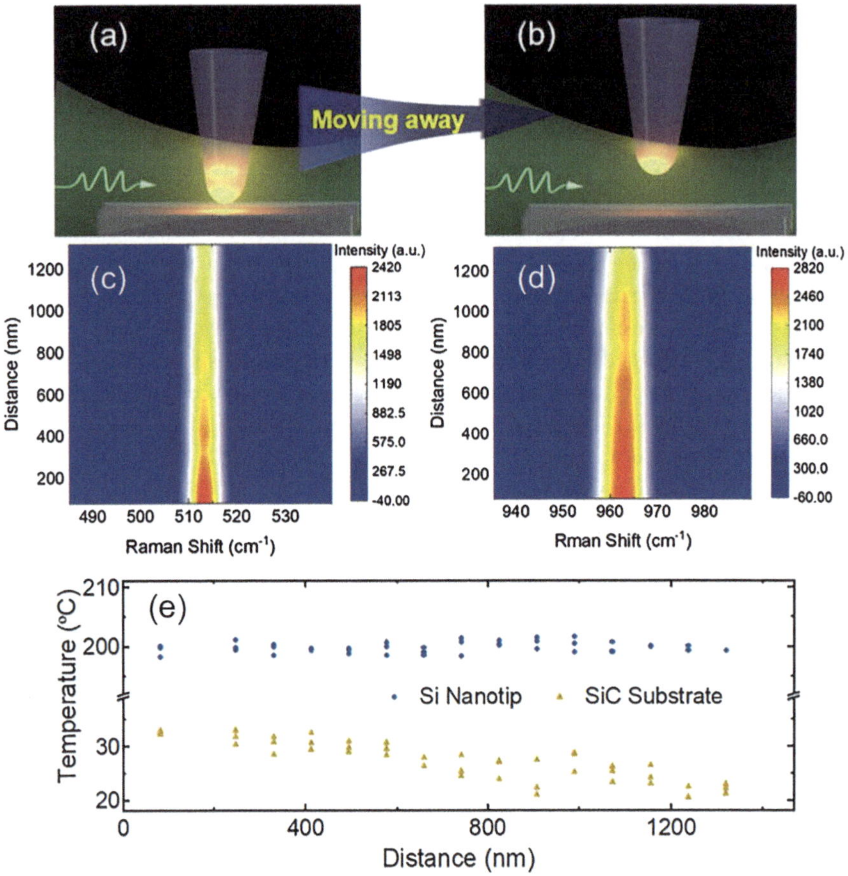

Fig. 9.3 The Si nanotip moving away from SiC substrate from (**a**) near contact mode to (**b**) a distance of 1320.0 nm. Raman scattering intensities of (**c**) Si nanotip and (**d**) SiC substrate at various clearances. (**e**) Temperatures of Si nanotip and SiC substrate at different gap sizes. (Reprinted from ref. [11], with permission from AIP Publishing)

localized temperature of the confined region. The measured decrease in substrate temperature indicates reduced energy absorption as the tip retracts, highlighting a pronounced enhancement in heat transfer at nanoscale separation. Temperature measurement uncertainties are derived from Raman shift fitting, with maximum deviations of ±0.026 and ±0.010 cm^{-1} for the nanotip and substrate, corresponding to temperature uncertainties of ±1.2 and ±0.4 °C, respectively.

Several factors can influence the substrate temperature, including direct laser irradiation, tip-enhanced electromagnetic field effects, and heat transfer from the heated nanotip. To clarify the dominant mechanism, additional optical experiments and electromagnetic simulations were conducted. Direct irradiation of the substrate with a 532 nm laser focused through a 50× objective produced no measurable temperature rise, consistent with the low absorptivity of SiC at this wavelength [13]. Moreover, finite element simulations based on three-dimensional Maxwell's equations revealed negligible electromagnetic field enhancement in the substrate, even at the smallest tip–substrate gap of 82.5 nm under experimental conditions. These results indicate that the observed substrate heating primarily arises from near-field heat transfer from the nanotip, which is strongly dependent on the temperature gradient and nanoscale separation. The elevated temperature enhances energy transfer between the tip and substrate, highlighting a significant nanoscale heat transfer effect. This phenomenon is likely associated with intensified air conduction and near-field radiative contributions, as supported by previous theoretical studies [4, 14]. The method introduced here provides a novel non-contact approach for directly measuring temperatures on both surfaces, significantly reducing uncertainty compared to conventional thermal-resistivity-based techniques.

Due to the non-flat geometry of the nanotip apex, the substrate region directly beneath the tip experiences the highest heating. With a nanoscale apex radius of ~10 nm, thermal transport in the heated SiC region is likely limited by ballistic effects, resulting in reduced effective thermal conductivity [15, 16]. Previous studies indicate that the thermal conductivity of Si and GaN can drop to ~10% of bulk values when heater sizes are below 50 and 100 nm, respectively [15]. The total heated area, however, can extend beyond the tip apex due to complex heat transfer phenomena in the surrounding region [17]. Thermal and vibrational fluctuations can cause minor probe drift, but in the experiments this drift is limited to a few nanometers, [18, 19] which has negligible effect on gap control. Hence, a minimum gap of 82.5 nm is employed, as maintaining smaller separations proves challenging. During measurements, the probe is held in a steady state, and its temperature remains nearly constant, further minimizing thermal drift and ensuring reliable gap control.

9.5 Heat Conduction Effect Induced by Air Molecules

The heat transfer mechanisms associated with the nanotip are depicted in Fig. 9.4a–c. These involve heat absorption by the nanotip (Q_{laser}), heat dissipation to the upper part (Q_{up}) through thermal conduction to the cantilever and thermal convection to the surroundings, and heat dissipation to the substrate (Q_{gap}) through both radiation and air conduction. As the nanotip moves away, its temperature remains relatively stable despite these intricate surrounding conditions, leading to a diminished heating effect on the substrate. To deepen the understanding of the thermal transport process, theoretical calculations are performed focusing on air conduction and radiation between the nanotip and substrate.

Given the sub-wavelength scale distance between the tip and substrate, heat transfer across the air gap is critically affected, especially when the clearance nears the mean free path of ambient air, resulting in a transition regime, and ballistic effect on thermal conductivity is considered. The air gap region was discretized into concentric cylindrical sections [3] with different gap distance, as depicted in Fig. 9.4c.

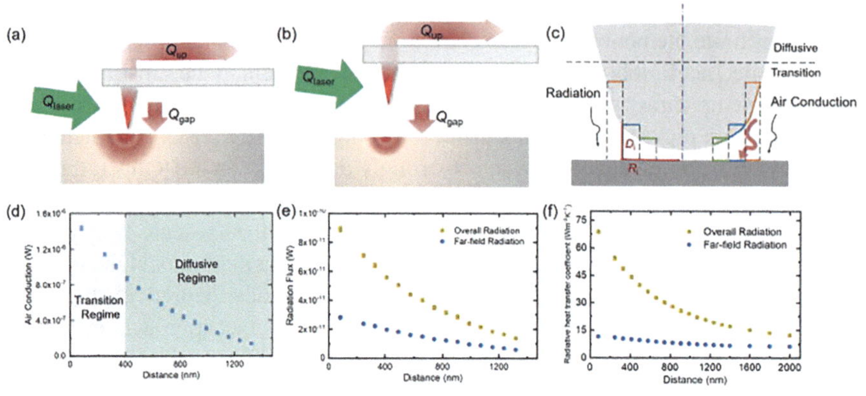

Fig. 9.4 Heat transfer in the nanotip–substrate system: (**a**) near contact, (**b**) large gap, (**c**) air gap decomposition for conduction and radiation calculations, (**d**) air conduction flux, (**e**) radiative flux, and (**f**) radiative heat transfer coefficient versus gap size. (Reprinted from ref. [11], with permission from AIP Publishing)

For gap clearance ≤ 400 nm, the heat conduction is treated as being in a transition regime. The effective thermal conductivity for each layer, based on its local clearance, is applied to calculate air conduction discretely. The air thermal conductivity is assumed to correspond to its bulk value for gap distance beyond 400 nm. Considering the minimal change in average air temperature during the experiment (less than 6 °C) and the small variation in air thermal conductivity (<1.1%), [20] the heat transfer enhancement due to temperature increase in the air is neglected. The computational analysis demonstrates an air conduction of 1.45×10^{-6} W at a gap distance of 82.5 nm and 1.38×10^{-7} W at 1320.0 nm, as shown in Fig. 9.4d. A significant reduction, by a factor of 9.51, in air conduction is observed as the tip moves away from the substrate, showing a substantial enhancement in air conduction as the gap distance decreases.

Near-field thermal radiation becomes significant when the gap distance shrinks to the sub-wavelength scale [21, 22]. To evaluate this effect, the spectral radiative heat flux is firstly computed at each angular frequency and then integrated from zero to infinity, with explicit consideration of the wavelength, temperature, and dielectric function of the interacting materials. The theoretical radiative flux is analyzed within the framework of fluctuational electrodynamics. Using the Derjaguin approximation [2] and a discretization method similar to that employed in the air conduction analysis, the local radiative fluxes between two parallel plates at different separations are calculated [2, 23] and subsequently integrated to yield the total radiative flux.

Figure 9.4e presents the theoretical radiative heat flux between the nanotip and substrate, which decreases monotonically as the tip retracts. The flux drops from 8.8×10^{-10} W at a gap of 82.5 nm to 9.6×10^{-11} W at 1320 nm, representing a 9.17-fold reduction. This trend mirrors that observed for air conduction, highlighting that radiation is significantly enhanced when the tip-substrate separation reaches tens of nanometers. Figure 9.4f shows the overall and far-field radiative heat transfer coefficients between parallel Si and SiC plates. At a gap distance of 82.5 nm and under experimental temperatures, the overall radiative heat transfer coefficient reaches 69.0 W/m^{2}K, whereas the far-field contribution is only 11.7 W/m^{2}K, emphasizing the dominant role of evanescent waves in near-field radiation. For the actual nanotip–substrate system, the enhancement in radiative flux is less pronounced because much of the nanotip lies farther from the substrate. In addition to the near-field effect, geometric factors such as the effective contact area and angular distribution of the tip also contribute to the increased radiative flux at small gaps. Despite this enhancement, the magnitude of radiative heat flux remains approximately three orders of magnitude smaller than that of air conduction, confirming that air-mediated heat transfer dominates the nanoscale thermal exchange between the nanotip and substrate, while radiative contributions are negligible.

9.6 Summary

This chapter explores near-field heat transfer, and the challenges it faces in experimental measurements. Based on Raman, experiments with a silicon nanotip and SiC substrate are conducted, showing enhanced heat transfer as gaps decrease. Developing in-situ Raman systems capable of operating under variable environmental conditions could further clarify the role of medium properties in near-field transport. These advancements may pave the way for applications in thermal rectifiers and nanoscale energy conversion devices.

References

1. Liu, X., Wang, L., Zhang, Z.M.: Near-field thermal radiation: recent progress and outlook. Nanoscale Microscale Thermophys. Eng. **19**(2), 98–126 (2015)
2. Polder, D., Vanhove, M.: Theory of radiative heat transfer between closely spaced bodies. Phys. Rev. B. **4**(10), 3303–3314 (1971)
3. Zhang, Y., Zhu, W.K., Borca-Tasciuc, T.: Thermal conductivity measurements of thin films by non-contact scanning thermal microscopy under ambient conditions. Nanoscale Adv. **3**(3), 692–702 (2021)
4. Zhang, Y., Zhu, W.K., Han, L., Borca-Tasciuc, T.: Quantitative temperature distribution measurements by non-contact scanning thermal microscopy using Wollaston probes under ambient conditions. Rev. Sci. Instrum. **91**(1), 014901 (2020)
5. Kim, K., Song, B., Fernandez-Hurtado, V., Lee, W., Jeong, W.H., Cui, L.J., Thompson, D., Feist, J., Reid, M.T.H., Garcia-Vidal, F.J., et al.: Radiative heat transfer in the extreme near field. Nature. **528**(7582), 387–391 (2015)
6. Salihoglu, H., Nam, W., Traverso, L., Segovia, M., Venuthurumilli, P.K., Liu, W., Wei, Y., Li, W.J., Xu, X.F.: Near-field thermal radiation between two plates with sub-10 nm vacuum separation. Nano Lett. **20**(8), 6091–6096 (2020)
7. DeSutter, J., Tang, L., Francoeur, M.: A near-field radiative heat transfer device. Nat. Nanotechnol. **14**(8), 751–755 (2019)

8. Luo, X., Salihoglu, H., Wang, Z.X., Li, Z., Kim, H., Liu, X., Li, J.Y., Yu, B.W., Du, S., Shen, S.: Observation of near-field thermal radiation between coplanar nanodevices with subwavelength dimensions. Nano Lett. **24**(5), 1502–1509 (2024)
9. Wang, R.D., Xu, S., Yue, Y.A., Wang, X.W.: Thermal behavior of materials in laser-assisted extreme manufacturing: Raman-based novel characterization. Int. J. Extreme Manuf. **2**(3), 032004 (2020)
10. Hamian, S., Yun, J., Park, I., Park, K.: Quantitative probing of tip-induced local cooling with a resistive nanoheater/thermometer. Appl. Phys. Lett. **109**(25), 253114 (2016)
11. Huang, X., Sun, Q., Xu, S., Yue, Y., Wang, X., Xuan, Y.: Observation of enhanced heat transfer between a nanotip and substrate at nanoscale distances via direct temperature probing with Raman spectroscopy. Appl. Phys. Lett. **125**(7), 072201 (2024)
12. Deng, S.G., Xu, S., Gao, J.S., Wu, H., She, J.J., Yue, Y.A.: Picosecond-resolved Raman response of a Si nanotip for probing temperature and thermal stress in the confined regime under laser heating. J. Phys. Chem. C. **126**(4), 1922–1930 (2022)
13. Ahuja, R., da Silva, A.F., Persson, C., Osorio-Guillén, J.M., Pepe, I., Järrendahl, K., Lindquist, O.P.A., Edwards, N.V., Wahab, Q., Johansson, B.: Optical properties of 4H-SiC. J. Appl. Phys. **91**(4), 2099–2103 (2002)
14. Fiorino, A., Zhu, L.X., Thompson, D., Mittapally, R., Reddy, P., Meyhofer, E.: Nanogap near-field thermophotovoltaics. Nat. Nanotechnol. **13**(9), 806–811 (2018)
15. Hu, Y.J., Zeng, L.P., Minnich, A.J., Dresselhaus, M.S., Chen, G.: Spectral mapping of thermal conductivity through nanoscale ballistic transport. Nat. Nanotechnol. **10**(8), 701–706 (2015)
16. Mahan, G.D., Claro, F.: Nonlocal theory of thermal-conductivity. Phys. Rev. B. **38**(3), 1963–1969 (1988)
17. Zhang, Y., Zhu, W.K., Hui, F., Lanza, M., Borca-Tasciuc, T., Rojo, M.M.: A review on principles and applications of scanning thermal microscopy (SThM). Adv. Funct. Mater. **30**(18), 1900892 (2020)
18. Okada, Y., Tokumaru, Y.: Precise determination of lattice-parameter and thermal-expansion coefficient of silicon between 300-K and 1500-K. J. Appl. Phys. **56**(2), 314–320 (1984)
19. Kato, R., Moriyama, T., Umakoshi, T., Yano, T.A., Verma, P.: Ultrastable tip-enhanced hyperspectral optical nanoimaging for defect analysis of large-sized WS2 layers. Sci. Adv. **8**(28), eabo4021 (2022)
20. Ochs, F., Heidemann, W., Müller-Steinhagen, H.: Effective thermal conductivity of moistened insulation materials as a function of temperature. Int. J. Heat Mass Transf. **51**(3–4), 539–552 (2008)
21. Mulet, J.P., Joulain, K., Carminati, R., Greffet, J.J.: Nanoscale radiative heat transfer between a small particle and a plane surface. Appl. Phys. Lett. **78**(19), 2931–2933 (2001)
22. Song, B., Ganjeh, Y., Sadat, S., Thompson, D., Fiorino, A., Fernández-Hurtado, V., Feist, J., Garcia-Vidal, F.J., Cuevas, J.C., Reddy, P., et al.: Enhancement of near-field radiative heat transfer using polar dielectric thin films. Nat. Nanotechnol. **10**(3), 253–258 (2015)
23. Tokunaga, T., Jarzembski, A., Shiga, T., Park, K., Francoeur, M.: Extreme near-field heat transfer between gold surfaces. Phys. Rev. B. **104**(12), 125404 (2021)

Chapter 10
Raman Thermometry in Bubble Generation and Energy Storage

Abstract Raman thermometry has expanded its applications across a diverse range of interdisciplinary fields due to its non-contact operation, nanoscale spatial resolution, and ability to correlate thermal signatures with material-specific vibrational modes. These unique capabilities enable the exploration of complex thermal dynamics in scenarios where conventional probes often fail to provide precise information. This chapter highlights emerging applications of Raman thermometry in bubble generation and energy storage. Section 10.1 discusses its use in monitoring and controlling the nucleation and growth of plasmonic bubbles, providing insights into localized heating and thermal dynamics at the nanoscale. Section 10.2 focuses on nanoparticle trapping for Raman detection, demonstrating how Raman-based techniques can capture thermal behaviour in micro- and nanoscale systems, including environmental applications such as the detection of microplastics. Section 10.3 explores Raman thermometry in thermochemical energy storage application.

Keywords Raman thermometry · Bubble generation · Nanoparticle trapping · Thermochemical energy storage

10.1 Raman Thermometry in Bubble Generation

Plasmonic bubbles have emerged as effective tools in diverse applications, including micro/nano-fabrication, energy conversion, and heat transfer enhancement. Their ability to concentrate heat at the nanoscale facilitates controlled phase transitions is crucial in complex systems such as biomedical platforms [1] and microfluidic devices [2]. Recent studies have already demonstrated that plasmonic nanoparticles (NPs), particularly Au NPs, convert optical energy into highly localized heat, thereby inducing the rapid nucleation of vapor bubbles. Here, a nanotip-Au NP hybrid system is introduced to reduce the laser power density required for bubble nucleation while enabling spatially precise bubble formation. To investigate the enhanced heating effect in the nanotip-Au NP system, a comparative experiment

Y. Yue, *Raman Thermometry*, https://doi.org/10.1007/978-981-95-5985-5_10

was conducted to measure the Raman thermal response of a silicon nanotip alone and in combination with Au NPs. It has been shown that the localized temperature of a silicon nanotip can rise significantly under laser irradiation in air, primarily due to the size-dependent heating effect [3].

Using Raman spectroscopy, a non-contact thermometry with high spatial resolution [4], temperatures of a bare nanotip and a nanotip decorated with Au NPs are measured. As shown in Fig. 10.1a, a same laser beam, simultaneously serving as heating source and exciting Raman scattering, is vertically focused on the nanotips through an objective lens. Temperature rises are typically determined through the calibration experiment and power experiment. Calibrated experiment aims at obtaining the correlation between intrinsic properties of Raman characteristic peaks and temperature, e.g., intensity ratio [5], full width at half maximum (FWHM) [6], and Raman shift [7]. While power experiment can quantify changes of the characteristic Raman peak under different powers. Due to the severe influence of thermal stress on Raman shift, here, FWHM is used to quantify the temperatures of nanotips.

Previous work has reported the parabolic relationship between FWHM and temperature [8], expressed as $\Gamma = B(T_\Gamma - T_0)^2 + C(T_\Gamma - T_0) + \Gamma_0$. Here, T_Γ is the unbiased FWHM-based temperature, B and C are calibration constants, and Γ_0 is the FWHM at room temperature. After conducting Lorentz peak-fitting on the Raman spectra of

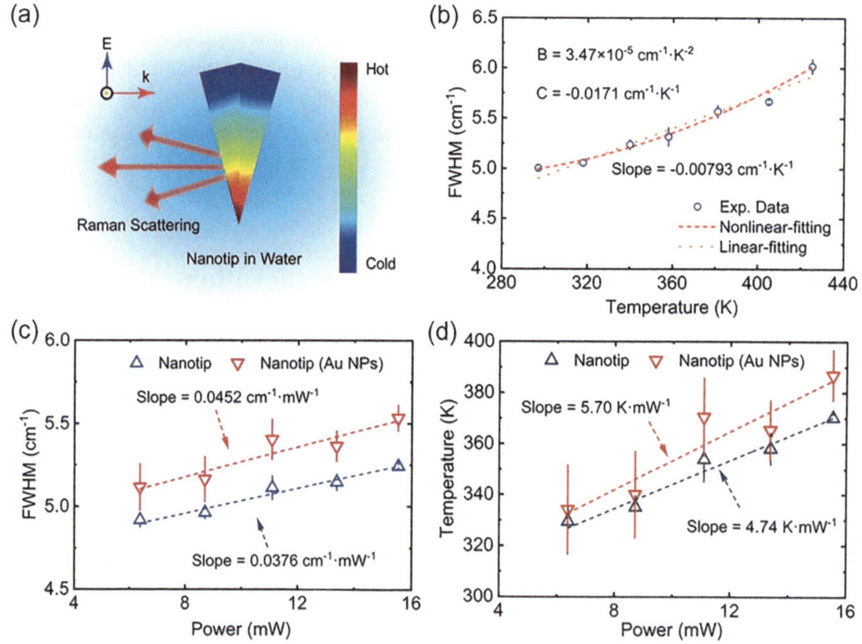

Fig. 10.1 (**a**) Schematic of laser heating on a nanotip in water. (**b**) Calibration with a standard AFM chip. (**c**) FWHM changes of the nanotip and nanotip–Au NP system under varying laser powers. (**d**) Linear relationship between laser power and nanotip temperature

silicon base, the relationship between the FWHM and temperature is shown in Fig. 10.1b. These data are firstly fitted with a quadratic function, showing that B and C are 3.47×10^{-5} cm^{-1}/K^2 and -0.0171 cm^{-1}/K, respectively. In the meantime, a linear-fitting result presents a slope of 0.00793 cm^{-1}/K, which is a good agreement with reported work [9]. Due to the proximity of the two fitted results, the more convenient linear fitting is selected as calibrated results. And the positively correlated trend results from changes in phonon lifetime as the temperature increases, and is further utilized to quantify local temperature of nanotip. Figure 10.1c shows the linear trends of FWHM and powers ranging from 6.38 mW to 15.53 mW, with slopes of 0.0376 cm^{-1}/mW for a bare nanotip and 0.0452 cm^{-1}/mW for a nanotip decorated with Au NPs, respectively. Taken these results together, the tips' temperatures under different laser powers are plotted in Fig. 10.1d, which finally shows the temperature-power coefficients of 4.74 K/mW and 5.70 K/mW after linear fitting. The growth rate of the coefficients is used to quantify the increased heat, expressed as $g = (tp_{Au} - tp_{tip})/tp_{tip}$, where tp represents the temperature-power coefficient. Finally, the calculated growth rate is 20%, verifying the enhanced heat in the nanotip–Au NPs compound system.

Tip-enhanced plasmonic bubble generation (TEPBG) technique was proposed to overcome the challenge of simultaneously monitoring plasmonic bubble nucleation and performing local Raman spectroscopy, a critical requirement for unraveling the interplay between plasmonic heating and bubble formation. As shown in Fig. 10.2a, in the setup, a silicon nanotip decorated with Au NPs, assembled via optical forces, is immersed in deionized water within a transparent quartz dish. A 532 nm continuous-wave laser, aligned with the tip's orientation to match the laser polarization and maximize plasmonic heating, is vertically focused onto the nanotip through a 20× objective lens; this laser not only heats the nanotip–Au NP system but also

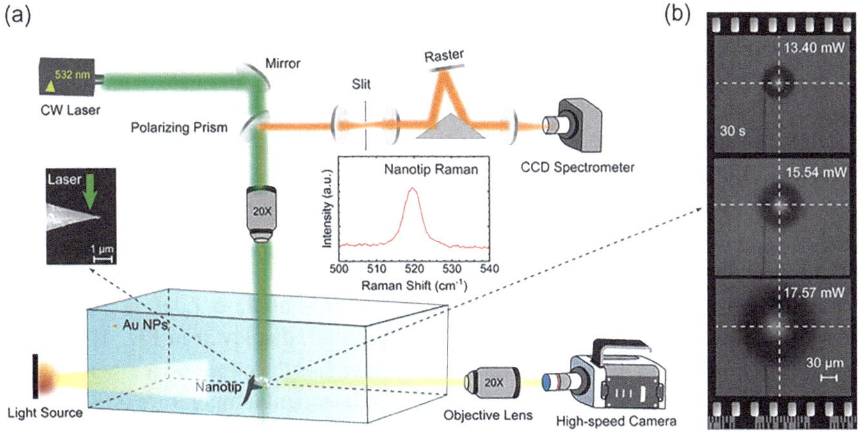

Fig. 10.2 Tip-enhanced plasmonic bubble generation (TEPBG) technique. (**a**) Schematic of the optical setup. (**b**) High-speed camera images of bubble nucleation in water obtained under laser powers from 13.40 mW to 17.57 mW

excites Raman signals near the nanotip. The resulting Raman spectra are collected by a cooled CCD detector and processed by integrated software, while a high-speed camera (also via a 20× objective) records the bubble nucleation dynamics under LED illumination.

By varying the laser power from 13.40 mW to 17.57 mW and observing for 30 s, displayed in Fig. 10.2b, it can be found that higher power produces larger bubbles due to a faster temperature rise and increased liquid overheating. Notably, the nanotip-Au NP assembly reduces the bubble nucleation threshold power density to 0.31 mW/μm^2, a 74% reduction compared to the conventional ~1.17 mW/μm^2 required when using dispersed Au NPs [10], demonstrating that the tip-enhanced plasmonic effect not only amplifies local electromagnetic heating but also facilitates bubble nucleation at substantially lower power. The process of bubble nucleation is also studied in different-concentration Au NPs solutions, and the results reveal that the bubble sizes can be controlled by adjusting the concentration of Au NPs. In short, this integrated approach addresses the difficulty of concurrently capturing transient bubble dynamics and localized spectroscopic information, thereby providing a robust platform for low-power plasmonic bubble nucleation.

10.2 Nanoparticle Trapping for Raman Detection

To extend the TEPBG method, microplastic detection was performed in aqueous systems, including self-prepared polystyrene (PS) bead solutions and environmental samples. Temperature-gradient–induced surface tension changes generate Marangoni flows during bubble nucleation, entrapping surrounding particles at the bubble surface, shown in Fig. 10.3a. These particles are subsequently deposited onto the nanotip via the shrinking surface bubble deposition (SSBD) technique [11]. The Au NPs on the nanotip further enhance Raman signals of the adsorbed particles through surface- and tip-enhanced Raman spectroscopy [12].

Here, monodisperse PS microparticles with a mean diameter of 5 μm are employed as an aqueous system to conduct experiment. The high-speed camera records images in the bubble growth and disappearance stages. These images clearly indicates that PS beads in surrounding microscale domain are attracted onto the nanotip and bubble surfaces via Marangoni convection in the bubble growth stage, and after bubble disappearance, there are still many particles remaining on the tip attributed to the flow induced by the shrinking bubble. In order to further illustrate this influence from Marangoni convection, a tip-bubble model is built with 800 × 800 nm region through finite element method, displayed in Fig. 10.3b. The tip's surface temperature is set as 220 °C [10], which is the common temperature for plasmonic bubble nucleation by Au NPs, the environment temperature is 20 °C, and temperature gradient of the surface tension of the gas-liquid interface is set as -8×10^{-5} kg/(s^2·K). The final simulation reveals that Marangoni convection exists around the bubble, and will attract the PS beads, which is consistent with the observed phenomena.

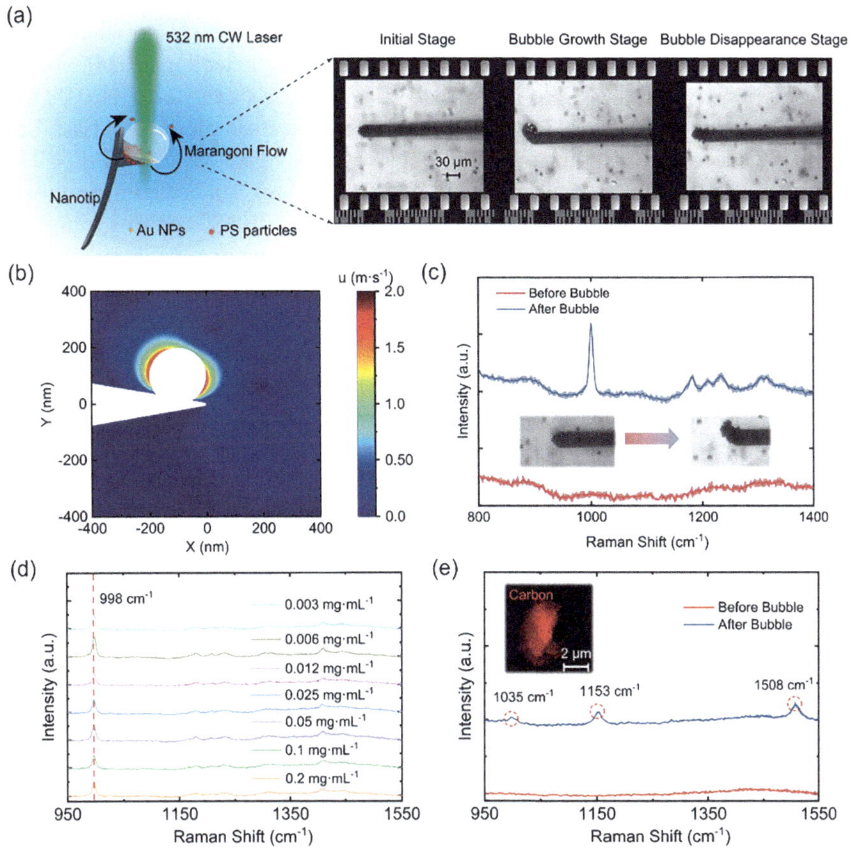

Fig. 10.3 (**a**) Raman probing of low-concentration PS solutions with high-speed images at different bubble stages. (**b**) Simulated velocity field around the bubble. (**c**) Raman signals near the nanotip before and after bubble formation. (**d**) Detection of micro-PS beads in dilute solutions. (**e**) Microplastic detection in natural water using TEPBG and EDS

To study the adsorption of particles on bubble surface dominated by Marangoni convection, Raman signals near the bubble are collected during stable bubble nucleation stage. Results indicate that bubble size has an obvious impact on Raman scattering intensity. In detail, Raman intensity shows a trend of first increasing and then decreasing with the increase of bubble radius. This phenomenon arises because the larger bubble will attract more particles, however, these particles are positioned beyond the laser focal plane. Although larger bubbles can attract more particles, the corresponding bubble contraction time will also significantly increase. Therefore, it is critical for detecting low-concentration substances to form appropriately sized bubbles.

Raman scattering near the nanotip was compared before bubble generation and after bubble disappearance in a 0.2 mg/mL PS solution, as shown in Fig. 10.3c.

Before bubble formation, the Raman signal was undetectable, showing that conventional SERS/TERS cannot sense such low concentrations. Using the TEPBG method, particles aggregated around the nanotip, enabling detection in 0.003–0.2 mg/mL solutions, which is shown in Fig. 10.3d, with clear signals even at 0.003 mg/mL. The method was then applied to environmental samples from East Lake, Wuhan, successfully detecting microplastics, as displayed in Fig. 10.3e. Characteristic Raman peaks at 1035, 1153, and 1508 cm^{-1} indicated mixed microplastic polymers, and Energy Dispersive Spectrometer (EDS) confirmed high carbon content, verifying anthropogenic microplastics. Notably, this approach allows low-power Raman measurements by moving the nanotip.

To sum up, a nanotip-Au NPs composite system reduces the laser power required for bubble nucleation by ~70% via enhanced local electromagnetic fields, with Raman thermometry quantifying thermal enhancement. This TEPBG method traps microplastics via Marangoni flow during bubble dynamics, depositing particles on the nanotip; Au NPs then amplify Raman signals for detection. The technique identifies polystyrene at concentrations as low as 0.003 mg/mL and detects microplastics in natural lake water, validated by energy-dispersive spectroscopy, showcasing Raman thermometry's ability in environmental monitoring.

10.3 Raman Thermometry in Energy Storage

The integration of thermochemical energy storage (TCES) with concentrated solar power (CSP) offers a solution to mitigate the inherent intermittency of solar irradiation by delivering high energy density and dispatchable operation [13–15]. TCES is based on reversible redox reactions, where thermal energy is stored in the form of chemical potential and released through exothermic back reactions, enabling stable utilization of high-temperature renewable heat [16–18]. Gas–solid metal oxide systems are particularly attractive, as air can function simultaneously as the heat transfer medium and the O_2 carrier, while their operating temperatures match those of CSP plants, thereby improving overall cycle efficiency [19–21]. Extensive studies have investigated various metal-oxide redox couples, including Co_3O_4/CoO, CuO/Cu_2O and Mn_2O_3/Mn_3O_4. Among them, the CuO/Cu_2O pair has gained prominence due to copper's earth abundance and the system's excellent redox reversibility.

However direct observation of nanoscale structural and compositional changes during thermochemical cycling remains highly challenging [22, 23]. Electron microscopy techniques such as field emission scanning electron microscopy and transmission electron microscopy offer sub-nanometer spatial resolution and can capture detailed morphologies, yet they deliver limited chemical information and cannot reproduce realistic gas–solid reaction environments due to stringent vacuum conditions and restricted operating regimes [24, 25]. Raman spectroscopy, in contrast, can provide compositional and defect-related insights by monitoring peak positions, intensities, and linewidths. Its ability to probe localized chemistry makes it particularly well suited for studying dynamic processes.

In this work, by using optical tweezers, individual CuO nanoparticles were manipulated and immobilized at the apex of a silicon nanotip. Building on this, we combined Raman spectroscopy with the localized photothermal enhancement of the nanotip to achieve simultaneous optical heating and high-sensitivity, in situ monitoring of nanoparticle transformations, thereby enabling real-time tracking of the thermochemical conversion from CuO to Cu_2O under precise spatial confinement and localized excitation. This method provides a robust platform for directly observing energy storage processes in nanostructured metal oxides and guides the design of high-performance nanoscale storage materials. To realize simultaneous in situ Raman detection with controlled photothermal activation of CuO nanoparticle reactions, we developed a confocal Raman-photothermal system, as shown in Fig. 10.4. A 532 nm continuous wave laser was used as the excitation source. The beam was conditioned through an adjustable expander and mirror array to optimize collimation and uniformity, followed by a band-pass filter to suppress stray light. Finally, the purified beam was focused by an objective lens onto the nanotip apex, generating a ~ 2 μm Gaussian spot for localized excitation.

As shown in Fig. 10.5, when the laser power increases from 15 mW to 34 mW, the characteristic CuO peak at ~296 cm^{-1} gradually intensifies, indicating stronger

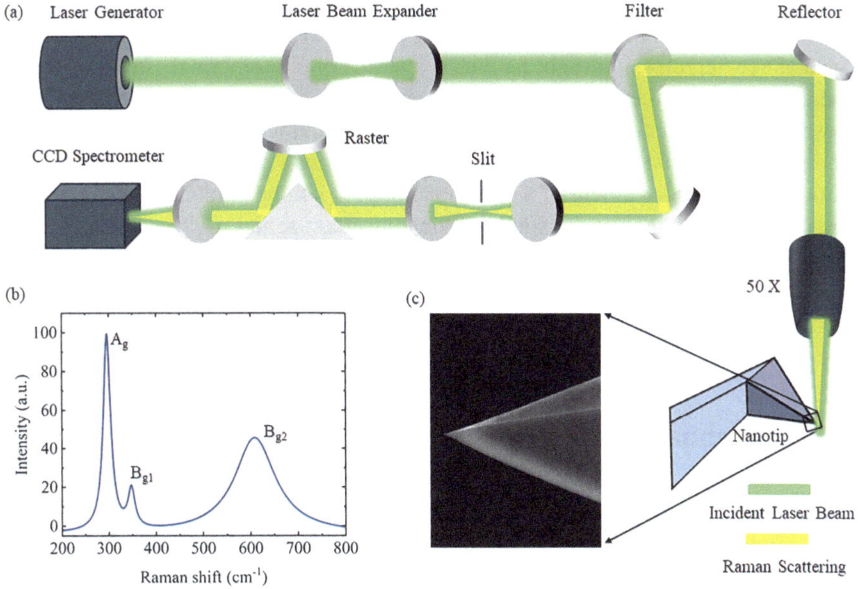

Fig. 10.4 Raman measurement setup and characterization of CuO nanoparticles. (**a**) Optical alignment for Raman exciation; (**b**) Typical Raman spectrum; (**c**) SEM and geometry pictures of the nanotip

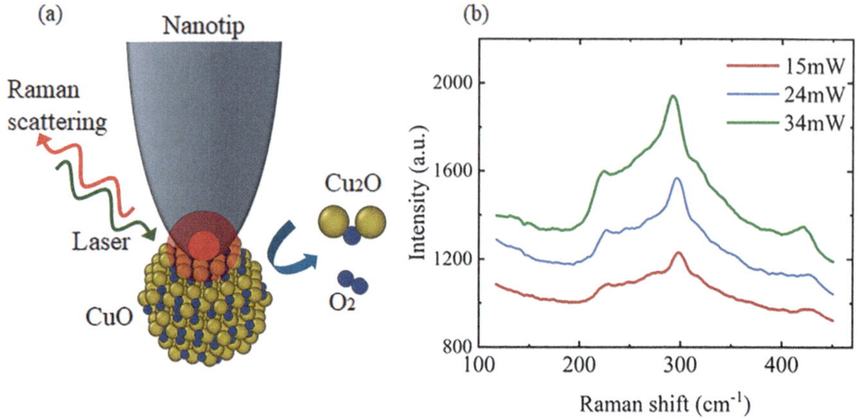

Fig. 10.5 Thermochemical reduction of individual CuO nanoparticles driven by laser irradiation. (**a**) Illustration of the attachment of the nanoparticle on the nanotip; (**b**) Characterized Raman spectra showing the signal of nanoparticle

lattice vibrations induced by localized photothermal effects. Meanwhile, a peak at ~226 cm^{-1} corresponding to Cu_2O emerges and grows in intensity, reflecting the partial reduction of the oxide. At powers of 24 mW and above, this product peak becomes more pronounced, signifying further progression of the reduction reaction. The spectra reveal not only enhanced peak intensities but also shifts and the appearance of new features, highlighting the critical role of photothermal effects in driving the CuO to Cu_2O thermochemical conversion. This section employs Raman spectroscopy to elucidate the nanoscale reduction behavior of the CuO/Cu_2O system under photothermal excitation. By leveraging optical tweezers and the localized photothermal enhancement of a silicon nanotip, in situ heating and real-time tracking of transformations at the single-particle level are achieved. The results demonstrate that laser power modulation can drive the stepwise conversion of CuO to Cu_2O, thereby confirming the critical role of photothermal effects in thermochemical energy storage reactions.

10.4 Summary

This chapter extends Raman thermometry beyond fundamental studies, focusing on practical applications like plasmonic bubble generation, microplastic detection, thermochemical energy storage. Future applications could include expanding TEPBG-based Raman thermometry to detect other trace contaminants, such as organic pollutants, by functionalizing nanoparticles for specific capture. In biomedicine, controlled plasmonic bubble generation combined with Raman monitoring may enable targeted drug delivery or precise thermal ablation in cancer therapy.

Integrating artificial intelligence with Raman spectral analysis could also automate pollutant identification, improving efficiency for large-scale environmental surveys.

References

1. Sarraf, S.S., Talabazar, F.R., Namli, I., Maleki, M., Aghdam, A.S., Gharib, G., Grishenkov, D., Ghorbani, M., Kosar, A.: Fundamentals, biomedical applications and future potential of micro-scale cavitation-a review. Lab Chip. **22**(12), 2237–2258 (2022)
2. Illath, K., Kar, S., Shinde, A., Ojha, R., Iyer, D.R., Mahapatra, N.R., Nagai, M., Santra, T.S.: Microfluidic device-fabricated spiky nano-burflower shape gold nanomaterials facilitate large biomolecule delivery into cells using infrared light pulses. Lab Chip. **23**(22), 4783–4803 (2023)
3. Deng, S., Xu, S., Gao, J., Wu, H., She, J., Yue, Y.: Picosecond-resolved Raman response of a Si nanotip for probing temperature and thermal stress in the confined regime under laser heating. J. Phys. Chem. C. **126**(4), 1922–1930 (2022)
4. Huang, D., Sun, Q., Liu, Z., Xu, S., Yang, R., Yue, Y.: Ballistic thermal transport at sub-10 nm laser-induced hot spots in GaN crystal. Adv. Sci. **10**(2), 2204777 (2023)
5. Sullivan, S., Vallabhaneni, A., Kholmanov, I., Ruan, X., Murthy, J., Shi, L.: Optical generation and detection of local nonequilibrium phonons in suspended graphene. Nano Lett. **17**(3), 2049–2056 (2017)
6. Sokalski, P., Han, Z., Fleming, G.C., Smith, B., Sullivan, S.E., Huang, R., Ruan, X., Shi, L.: Effects of hot phonons and thermal stress in micro-Raman spectra of molybdenum disulfide. Appl. Phys. Lett. **121**(18), 182202 (2022)
7. Balandin, A.A., Ghosh, S., Bao, W., Calizo, I., Teweldebrhan, D., Miao, F., Lau, C.N.: Superior thermal conductivity of single-layer graphene. Nano Lett. **8**(3), 902–907 (2008)
8. Beechem, T., Graham, S., Kearney, S.P., Phinney, L.M., Serrano, J.R.: Invited article: simultaneous mapping of temperature and stress in microdevices using micro-Raman spectroscopy. Rev. Sci. Instrum. **78**(6), 061301 (2007)
9. Su, Z., Sha, J., Pan, G., Liu, J., Yang, D., Dickinson, C., Zhou, W.: Temperature-dependent Raman scattering of silicon nanowires. J. Phys. Chem. B. **110**(3), 1229–1234 (2006)
10. Baffou, G., Polleux, J., Rigneault, H., Monneret, S.: Super-heating and micro-bubble generation around plasmonic nanoparticles under cw illumination. J. Phys. Chem. C. **118**(9), 4890–4898 (2014)
11. Moon, S., Martin, L.M.A., Kim, S., Zhang, Q., Zhang, R., Xu, W., Luo, T.: Direct observation and identification of nanoplastics in ocean water. Sci. Adv. **10**(4), eadh1675 (2024)
12. Höppener, C., Aizpurua, J., Chen, H., Gräfe, S., Jorio, A., Kupfer, S., Zhang, Z., Deckert, V.: Tip-enhanced Raman scattering. Nat. Rev. Methods Primers. **4**(1), 47 (2024)
13. Dizaji, H.B., Hosseini, H.: A review of material screening in pure and mixed-metal oxide thermochemical energy storage (TCES) systems for concentrated solar power (CSP) applications. Renew. Sust. Energ. Rev. **98**, 9–26 (2018)
14. Kennedy, K.M., Ruggles, T.H., Rinaldi, K., Dowling, J.A., Duan, L., Caldeira, K., Lewis, N.S.: The role of concentrated solar power with thermal energy storage in least-cost highly reliable electricity systems fully powered by variable renewable energy. Adv. Appl. Energ. **6**, 100091 (2022)
15. Wu, R., Deng, L., Luo, X., Zheng, H., Lu, P., Liu, L., Huang, H., Chihara, N., Kubota, M., Kobayashi, N.: Development of LiCoO2 and CuO co-doped cobalt oxide composites for the comprehensive performance improvement of thermochemical energy storage in concentrated solar power plants. Energy. **327**, 136435 (2025)
16. Sunku Prasad, J., Muthukumar, P., Desai, F., Basu, D.N., Rahman, M.M.: A critical review of high-temperature reversible thermochemical energy storage systems. Appl. Energy. **254**, 113733 (2019)

17. Salgado-Pizarro, R., Calderón, A., Svobodova-Sedlackova, A., Fernández, A.I., Barreneche, C.: The relevance of thermochemical energy storage in the last two decades: the analysis of research evolution. J. Energy Storage. **51**, 104377 (2022)
18. Ali, H.M., Rehman, T.-u., Arıcı, M., Said, Z., Duraković, B., Mohammed, H.I., Kumar, R., Rathod, M.K., Buyukdagli, O., Teggar, M.: Advances in thermal energy storage: fundamentals and applications. Prog. Energy Combust. Sci. **100**, 101109 (2024)
19. Schrader, A.J., Bush, H.E., Ranjan, D., Loutzenhiser, P.G.: Aluminum-doped calcium manganite particles for solar thermochemical energy storage: reactor design, particle characterization, and heat and mass transfer modeling. Int. J. Heat Mass Transf. **152**, 119461 (2020)
20. Zhao, Y., Zhao, C.Y., Markides, C.N., Wang, H., Li, W.: Medium- and high-temperature latent and thermochemical heat storage using metals and metallic compounds as heat storage media: a technical review. Appl. Energy. **280**, 115950 (2020)
21. Deng, J., Gu, C., Xu, H., Xiao, G.: MgCr(2)O(4)-modified CuO/Cu(2)O for high-temperature thermochemical energy storage with high redox activity and sintering resistance. ACS Appl. Mater. Interfaces. **14**(38), 43151–43162 (2022)
22. Liu, J., Baeyens, J., Deng, Y., Wang, X., Zhang, H.: High temperature Mn(2)O(3)/Mn(3)O(4) and Co(3)O(4)/CoO systems for thermo-chemical energy storage. J. Environ. Manag. **267**, 110582 (2020)
23. Wei, J., Wang, M., Tang, G., Akhtar, M.A., Xu, D., Song, X., Yu, G., Li, B., Zhang, H., Zhang, S.: Advances on in-situ analysis of char structure evolution during thermochemical conversion of coal/biomass: a review. Fuel Process. Technol. **230**, 107221 (2022)
24. Lopez, L.E.P., Roslawska, A., Scheurer, F., Berciaud, S., Schull, G.: Tip-induced excitonic luminescence nanoscopy of an atomically resolved van der Waals heterostructure. Nat. Mater. **22**(4), 482–488 (2023)
25. Li, L., Wang, M., Zhou, Y., Zhang, Y., Zhang, F., Wu, Y., Wang, Y., Lyu, Y., Lu, N., Wang, G., et al.: Manipulating the insulator-metal transition through tip-induced hydrogenation. Nat. Mater. **21**(11), 1246–1251 (2022)

Chapter 11
Challenges and Opportunities

Abstract Raman thermometry has become an indispensable tool for probing thermal phenomena at micro- and nanoscale dimensions, yet several challenges remain that limit its broader application and quantitative accuracy. Section 11.1 discusses the technical challenges in Raman-based thermal measurements. Section 11.2 explores future opportunities for Raman measurement, not just thermometry, but also with biomedical application, manufacturing application, as well as AI-based Raman characterizations.

Keywords Raman thermometry · Challenges · Future applications · Biomedical · Manufacturing · AI-based Raman characterization

11.1 Challenges in Raman Thermometry

Despite the significant advancements in Raman thermometry and its growing applications across nanotechnology, material science, and semiconductor research, the technique remains inherently constrained by fundamental limitations [1–3]. Foremost among these is its applicability only to Raman-active materials. Raman thermometry relies on the inelastic scattering of photons, and as such, it requires materials with vibrational modes that can effectively interact with the incident light [1]. Unfortunately, a substantial number of materials, including metals and amorphous solids, are either Raman inactive or exhibit extremely weak Raman scattering, thereby rendering the ineffective or entirely unusable in such contexts.

A second major limitation lies in the inherently weak nature of the Raman signal itself. For most materials, the Raman scattering cross-section is several orders of magnitude lower than that of other optical processes such as fluorescence or elastic (Rayleigh) scattering. This requires the use of highly sensitive detectors, long integration times, and often high-power excitation sources to achieve a measurable signal-to-noise ratio. These requirements introduce challenges not only in terms of instrumentation costs and measurement time but also raise practical issues when

Y. Yue, *Raman Thermometry*, https://doi.org/10.1007/978-981-95-5985-5_11

dealing with low-dimensional systems, where the volume of Raman-active material is minimal and thus the scattered signal is further diminished [1].

Compounding these limitations is the persistent issue of laser-induced heating during the Raman measurement process [3, 4]. Since Raman thermometry involves using a focused laser beam to probe temperature-dependent shifts in vibrational modes, the very act of probing can alter the thermal equilibrium of the system. This is particularly problematic when dealing with soft materials, such as polymers, organics, and biological specimens, which often exhibit low thermal degradation or phase transition thresholds [2]. Even moderate laser powers may induce localized heating that significantly impacts the measured temperature or damages the sample [5].

Moreover, the interpretation of temperature from Raman spectra is itself not always straightforward [6]. The correlation between Raman peak position, linewidth, and intensity ratio with temperature can be affected by a range of confounding factors such as strain, doping, and crystallinity, especially in heterogeneous or nanoscale systems. Excluding these effects to purely thermal contributions require careful calibration, often under controlled environments that may not reflect actual device operating conditions. This complexity introduces further uncertainty that must be considered in quantitative analyses.

Another key challenge lies in the spatial and temporal resolution limits of Raman thermometry. While advances in confocal and TERT technique have improved spatial resolution [5, 6], they are often constrained by the diffraction limit and the necessity for long acquisition times to accumulate sufficient signal. As a result, real-time thermal mapping at the nanoscale, especially under dynamic conditions such as optical excitation, remains a difficult task. This limits the applicability of Raman thermometry in high-speed or transient thermal environments where rapid thermal fluctuations occur [7–9].

Lastly, the integration of Raman thermometry into complex experimental platforms or real-world devices poses additional practical difficulties. Factors such as optical access, environmental stability, sample preparation, and alignment precision can all affect the feasibility and reliability of Raman-based thermal measurements. For in situ studies, particularly under high-pressure, high-temperature, or reactive environments, these logistical challenges can significantly limit the applicability of the technique [10–14]. Addressing these issues will require ongoing innovation in instrumentation design, data interpretation methodologies, and material-specific calibration protocols.

11.2 Future Opportunities for Applications

11.2.1 Raman Spectroscopy in Biomedical Applications

While Raman faces intrinsic limitations in certain materials and operational environments, significant opportunities remain. One of the most promising areas is in biomedical diagnostics, where Raman spectroscopy has shown great potential for label-free, non-invasive cancer detection. The technique enables molecular fingerprinting of biological tissues with high specificity, making it capable of distinguishing between healthy and malignant cells based on subtle biochemical differences in their vibrational spectra. This opens pathways for early cancer diagnosis, intraoperative margin assessment, and even real-time in vivo screening, especially when coupled with fiber-optics.

Beyond diagnostics, Raman-based methods are also being explored for photothermal cancer therapy, where they serve a dual role in theranostics: combining therapy and diagnostics in a single platform [15]. In this approach, plasmonic nanoparticles (e.g., gold nanorods or shells) are functionalized to target tumor cells and are then irradiated with near-infrared laser light. Raman spectroscopy can be used to monitor the thermal effects in situ by tracking temperature-sensitive Raman shifts, while simultaneously guiding and optimizing the photothermal treatment [16–18]. This real-time thermal feedback helps to minimize damage to surrounding healthy tissue and enhances treatment efficacy, making Raman thermometry an attractive tool for precision oncology.

In addition, the integration of Raman spectroscopy with machine learning and advanced imaging techniques is expanding its utility in biomedical research. Algorithms trained on large datasets of Raman spectra can accurately classify cell types, disease states, or tissue composition, thereby enhancing diagnostic confidence. Moreover, developments in stimulated Raman scattering (SRS) and coherent anti-Stokes Raman scattering (CARS) are enabling rapid, high-resolution imaging of biological samples with minimal photodamage. These innovations are steadily transforming Raman techniques from a niche research tool into a clinically relevant modality, offering both scientific insight and translational potential across medicine and life sciences.

11.2.2 Raman Spectroscopy in Advanced Manufacturing

Raman spectroscopy is increasingly being adopted in advanced manufacturing as a powerful, non-destructive analytical tool, particularly for in situ monitoring of battery materials [19]. During the operation of lithium-ion batteries, for example, Raman spectroscopy can provide real-time insight into phase transitions, lithium intercalation, solid electrolyte interphase (SEI) formation, and thermal behaviour of electrodes. These measurements are critical for understanding degradation mechanisms and optimizing performance. Unlike traditional electrochemical characterization techniques, Raman offers spatially resolved chemical information, making it ideal for tracking local changes within electrode materials under actual operating conditions [20]. This capability enables the development of smarter, safer batteries through improved design and real-time quality control during manufacturing.

In the field of welding and additive manufacturing, Raman spectroscopy is emerging as a novel technique for detecting defects and monitoring process quality in real time. Conventional inspection methods such as X-ray imaging or ultrasonic testing are often applied post-process and may not capture subtle chemical or phase-related issues that can lead to long-term structural failures. Raman spectroscopy, on the other hand, can detect microstructural changes, residual stresses, and phase transformations at the weld zone with high sensitivity. Its ability to identify oxide inclusions, contamination, or undesirable alloy phases during or immediately after the welding process provides manufacturers with critical feedback that can be used to adjust parameters on-the-fly, significantly enhancing process reliability.

Furthermore, the integration of Raman spectroscopy with automated manufacturing systems and machine learning algorithms is unlocking new levels of process intelligence. Embedded Raman sensors in production lines can enable continuous, high-throughput inspection of materials, feeding data into predictive models that anticipate failure modes or quality deviations before they occur. This aligns well with the goals of Industry 4.0 and smart manufacturing, where real-time data acquisition and adaptive control systems are essential. As the spatial and temporal resolution of Raman systems improve, and as compact, fiber-coupled or handheld Raman devices become more robust, the technique is poised to become a standard tool for in situ quality assurance and process optimization across a wide range of high-value manufacturing applications.

11.2.3 AI-assisted Raman Thermometry

The integration of artificial intelligence (AI) and machine learning (ML) into Raman thermometry presents a transformative opportunity to overcome longstanding challenges in spectral analysis and temperature extraction [21–23]. Traditional Raman data processing often relies on manual or semi-automated peak fitting routines, which can be sensitive to noise, baseline drift, and overlapping features—especially in low signal-to-noise conditions or complex materials systems. AI algorithms, particularly those based on supervised learning or deep learning, can be trained to automatically detect Raman peaks with high accuracy. These models can differentiate between temperature-induced peak shifts and other perturbations such as strain or compositional variations, thus enabling more robust and automated thermal analysis.

In addition to peak detection, AI significantly enhances spectrum profile integration by facilitating real-time deconvolution of overlapping vibrational modes and dynamic baseline correction [23–25]. This is particularly useful for temperature detection and peak fitting, where precise integration of spectral regions is crucial. AI-based spectral regression techniques can also predict temperature directly from raw spectra without the need for extensive pre-processing, offering a faster and potentially more accurate alternative to conventional fitting-based approaches. As Raman thermometry continues to be applied in complex, real-time environments, such as in battery studies or high-throughput manufacturing, AI-assisted analysis will be essential for enabling scalable, real-time thermal probing with minimal human intervention.

References

1. Yue, Y., Wang, X.: Nanoscale thermal probing. Nano Reviews. **3**, 11586 (2012)
2. Yue, Y., Eres, G., Wang, X., Guo, L.: Characterization of thermal transport in micro/nanoscale wires by steady-state electro-Raman-thermal technique. Appl. Phys. A Mater. Sci. Process. **97**, 19–23 (2009). https://doi.org/10.1007/s00339-009-5352-6
3. Wang, T., Xu, S., Hurley, D.H., Yue, Y., Wang, X.: Frequency-resolved Raman for transient thermal probing and thermal diffusivity measurement. Opt. Lett. **40**(20), 4823–4826 (2015). https://doi.org/10.1364/OL.40.004823
4. Xu, S., Wang, T., Hurley, D.H., Yue, Y., Wang, X.: Development of time-domain differential Raman for transient thermal probing of materials. Opt. Express. **23**(18), 23477–23486 (2015). https://doi.org/10.1364/OE.23.023477
5. Huang, D., Sun, Q., Liu, Z., Xu, S., Yang, R., Yue, Y.: Ballistic thermal transport at Sub-10 nm laser-induced hot spots in GaN crystal. Adv. Sci. **10**, 2206672 (2023). https://doi.org/10.1002/advs.202206672
6. J Xu, X Huang, Y Sheng, Q Sun, H Zhang, H Bao, Y Yue: Direct observation of substantial phonon nonequilibrium near nanoscale hotspots in gallium nitride. Adv. Sci. **12**(11), 2411040 (2025)
7. Yue, Y., Hu, X., et al.: In situ two-step Raman thermometry for thermal characterization of monolayer graphene interface material. Appl. Therm. Eng. **113**, 1434–1440 (2017). https://doi.org/10.1016/j.applthermaleng.2016.11.111
8. Li, C., Xu, S., Yue, Y., Yang, B., Wang, X.: Thermal characterization of carbon nanotube fiber by time-domain differential Raman. Carbon. **103**, 179–185 (2016). https://doi.org/10.1016/j.carbon.2016.03.026
9. Dong, B., Cheng, C., Liu, H.: Effect of laser-induced heating on Raman measurement within a silicon microfluidic channel. Micromachines. **6**, 813–826 (2015). https://doi.org/10.3390/mi6070813
10. Zhang, R., Zhang, Y., Dong, Z.C., et al.: Chemical mapping of a single molecule by Plasmon-enhanced Raman scattering. Nature. **498**, 82–86 (2013). https://doi.org/10.1038/nature12151
11. Pettinger, B., Schambach, P., Villagómez, C.J., Scott, N.: Tip-enhanced Raman spectroscopy: near-fields acting on a few molecules. Annu. Rev. Phys. Chem. **63**, 379–399 (2012). https://doi.org/10.1146/annurev-physchem-032511-143807
12. Kumar, N., Mignuzzi, S., Suh, J., et al.: Tip-enhanced Raman spectroscopy: challenges and recent advances. J. Phys. Chem. C. **125**(31), 17100–17117 (2021). https://doi.org/10.1021/acs.jpcc.1c05353
13. Stadler, J., Schmid, T., Zenobi, R.: Nanoscale chemical imaging using top-illumination tip-enhanced Raman spectroscopy. Nano Lett. **10**(11), 4514–4520 (2010). https://doi.org/10.1021/nl1023369
14. Zhang, Y., Zhang, S., Dong, J., et al.: Quantum limit in subnanometre-gap tip-enhanced nano-imaging of few-layer MoS_2. Nat. Photonics. **10**, 493–498 (2016). https://doi.org/10.1038/nphoton.2016.111
15. Ngo, D.N., Xuan, V.T.T., Kim, G., Song, M.S., Kim, M.-R., Choo, J., Joo, S.-W., Lee, S.Y.: Raman thermometry nanopipettes in cancer photothermal therapy. Anal. Chem. **94**(17), 6463–6472 (2022). https://doi.org/10.1021/acs.analchem.1c04452
16. Singh, V., Kumar, R., Rai, V., et al.: In vitro Raman thermometry using gold Nanorod-decorated carbon nanotubes. ACS Appl. Nano Mater. **7**(2), 1453–1461 (2024). https://doi.org/10.1021/acsanm.4c03865
17. Fukui, Y., Tamada, M., Watanabe, T., et al.: Raman microspectroscopic mapping of tumor ablation effect by photothermal therapy. Sci. Rep. **11**, 14673 (2021). https://doi.org/10.1038/s41598-021-94286-w
18. Iverson, N.M., Barone, P.W., Shandell, M., et al.: In vivo biosensing via tissue-localizable near-infrared fluorescent single-walled carbon nanotubes. Nat. Nanotechnol. **8**, 873–880 (2013). https://doi.org/10.1038/nnano.2013.222

19. Thermo Fisher Scientific: Raman and FTIR Solutions for Battery Research. https://www.thermofisher.com/us/en/home/materials-science/battery-research/technologies/raman-ftir.html. Accessed Oct 2025
20. Argonne National Laboratory: Raman Spectroscopy for In Situ Monitoring of Electrode Processes. ANL Report (1982). https://www.osti.gov/biblio/5292328
21. Zhang, Y., Jiang, M., Li, Y., et al.: An intuitive approach for spike removal in Raman spectra based on peaks' prominence and width. Anal. Chim. Acta. **1260**, 341413 (2024). https://doi.org/10.1016/j.aca.2023.341413
22. Qi, L., Fu, Y., Shi, Z., et al.: A new peak detection algorithm of Raman spectra based on bi-scale correlation. J. Biomed. Opt. **19**(2), 027006 (2014). https://doi.org/10.1117/1.JBO.19.2.027006
23. Li, B., Liu, M., Fang, Y., et al.: Automated peak recognition in Raman spectra based on wavelet transform and local SNR. Spectrochim. Acta A. **181**, 124–130 (2017). https://doi.org/10.1016/j.saa.2017.03.036
24. Yan, L., He, Y., Song, W., et al.: Feature Visualization of Raman Spectrum Analysis with Deep CNN. arXiv 2020, arXiv:2007.13354. https://arxiv.org/abs/2007.13354
25. Qi, J., Xie, H., Wu, Y., et al.: Identifying surface-enhanced Raman spectra with a Raman library using machine learning. ACS Nano. **17**(10), 9961–9971 (2023). https://doi.org/10.1021/acsnano.3c05510

GPSR Compliance

The European Union's (EU) General Product Safety Regulation (GPSR) is a set of rules that requires consumer products to be safe and our obligations to ensure this.

If you have any concerns about our products, you can contact us on ProductSafety@springernature.com

In case Publisher is established outside the EU, the EU authorized representative is:

Springer Nature Customer Service Center GmbH
Europaplatz 3
69115 Heidelberg, Germany

Batch number: 10755332

Printed by Printforce, the Netherlands